D1503534

INSTITUTE
OF
TRANSPORTATION ENGINEERS

Residential Street Design and Traffic Control

WOLFGANG S. HOMBURGER

*Institute of Transportation Studies,
and Department of Civil Engineering,
University of California, Berkeley*

ELIZABETH A. DEAKIN

*Institute of Transportation Studies,
and Department of City & Regional
Planning, University of California,
Berkeley*

PETER C. BOSSELMANN

*Institute of Urban & Regional
Development, and Department of City
& Regional Planning, University of
California, Berkeley*

DANIEL T. SMITH, JR.

DKS Associates, Oakland, CA

BERT BEUKERS

*Rijkswaterstaat, Transportation &
Traffic Engineering Division, Government
of the Netherlands, The Hague*

 PRENTICE HALL, Englewood Cliffs, New Jersey 07632

Library of Congress Cataloging-in-Publication Data

Residential street design and traffic control.

    Bibliography
    Includes index.
    1. Traffic engineering.   2. Roads—Design.
3. Neighborhood.   I. Homburger, Wolfgang S.
II Institute of Transportation Engineers.
HE335.R47   1989   625.7'25   88-17910
ISBN 0-13-775008-0

Editorial/production supervision and
   interior design: Carolyn Fellows and Joseph Scordato
Cover design: Edsal Enterprises
Manufacturing buyer: Mary Noonan

INSTITUTE OF TRANSPORTATION ENGINEERS
BOOKS AVAILABLE FROM PRENTICE HALL

Residential Street Design and Traffic Control

Traffic Signal Maintenance Manual

Transportation and Traffic Engineering Handbook

Transportation and Land Development

©1989 by Prentice-Hall, Inc.
A Division of Simon & Schuster
Englewood Cliffs, New Jersey 07632

The Institute of Transportation Engineers is an international
society of transportation professionals responsible for ensuring that surface transportation systems allow for the safe, efficient, and environmentally sound movement of people and
goods. Since 1930, ITE has been providing transportation
professionals with the programs and resources they need to
help them meet those responsibilities. ITE programs and
resources include publications, seminars, local, regional, and
international meetings, and other forums that allow for the ex-
change of opinions, ideas, techniques, and research in transportation engineering.

The Institute's more than 9,000 members are transportation engineers and planners working for highway, transit, port, and
terminal systems and facilities in the United States, Canada,
and 70 other countries.

Printed in the United States of America
10 9 8 7 6 5 4 3 2 1

INSTITUTE OF TRANSPORTATION ENGINEERS

525 School St., SW, Suite 410
Washington, DC 20024-2729 USA
Telephone: 202/554-8050
Telex: 467943 ITE WSH CI.
Fax: 202/863-5486

ISBN 0-13-775008-0

Prentice-Hall International (UK) Limited, *London*
Prentice-Hall of Australia Pty. Limited, *Sydney*
Prentice-Hall Canada Inc., *Toronto*
Prentice-Hall Hispanoamericana, S.A., *Mexico*
Prentice-Hall of India Private Limited, *New Delhi*
Prentice-Hall of Japan, Inc., *Tokyo*
Simon & Schuster Asia Pte. Ltd., *Singapore*
Editora Prentice-Hall do Brasil, Ltda., *Rio de Janeiro*

Contents

CHAPTER 6   ***IMPLEMENTING NEIGHBORHOOD TRAFFIC CONTROLS***   ***119***

CHAPTER 7   ***PROSPECTS***   ***134***

APPENDIX   ***136***

Preface

As the professional society for transportation engineers and planners responsible for ensuring that streets operate safely, efficiently, and in harmony with their environment, the Institute of Transportation Engineers recognized the need for providing transportation professionals and community leaders with strategies and techniques for creating compatible relationships between residential neighborhoods and streets. This book is an attempt to meet that need.

Residential Street Design and Traffic Control is intended for transportation engineers, urban planners, community groups, elected officials, and others concerned with the safety of residential streets and neighborhoods. It is a tool for understanding and finding solutions to the problems of noise and traffic control on those streets.

The authors bring to this book a wealth of knowledge and expertise in the design and traffic control of residential streets.

1
Introduction

Streets are public spaces for many activities and functions. They permit the diffusion of light and circulation of air. They offer opportunities for providing landscaped vistas, trees and shrubs, paths for walking, places for talking, rights-of-way for utilities, and—among all these other activities—facility for the movement, stopping, and storage of motor vehicles. Some streets open up into squares, others narrow into alleys. Some achieve fame or notoriety, most are known only to their residents and those who pass through regularly.

The first function of residential streets is to serve the land that abuts them. They provide for access to homes by all who enter and leave, and all who deliver and collect. But they are also routes for those who wish only to pass through the area. It is here that conflict arises, for there is a basic discrepancy between the impact of vehicular traffic and the tranquility of a residential street. The inherent kinetic energy of a moving vehicle produces noise, and contact with a hapless person, a fixed object, or another vehicle may cause a disastrous dissipation of that energy. If propelled by an internal combustion engine, the vehicle emits noxious and polluting fumes. If parked, it takes up space and blocks views and sight lines.

The street might well be a place not only for walking and mechanical transportation, but also for social activities: outdoor play of young children in front of their homes, the elderly sunning themselves on minipark benches, residents crossing to drop in on neighbors, joggers and cyclists partaking of healthful exercise. All these and more activities are imperiled by fast or heavy traffic streams.

Three specific forms of "unwanted traffic" are recognized on residential streets:

- Traffic using the streets as shortcuts, detours (such as going around a residential block because of a left-turn prohibition on an abutting major street), or overflow from a congested arterial.
- Excessive traffic speeds.
- Use of curb parking spaces (with related vehicle movements in search for and leaving such spaces) by drivers whose destinations are outside the neighborhood.

Often, such intrusion is possible because the geometry of street networks and of individual streets was fixed long before such conflicts had been visualized. But even "modern" residential subdivision streets may experience neighborhood traffic problems. While newer designs may have mitigated the problem of unwanted through traffic, they have, with their generous radii of curvature and ample parking lanes, done little to reduce problems of speeders and unwanted parkers.

The pervasive conflict between traffic and residential uses has been tolerated to a greater or lesser degree throughout the world. However, demands of residents for amelioration of their environment have been growing; since 1960, considerable literature in the planning and transportation fields has been devoted to analyzing the problem and outlining possible solutions.

This book sets forth the basic issues and goals involved in designing and operating circulation systems in residential neighborhoods to provide a more compatible relationship between home and traffic. It covers the planning principles and processes involved, discusses alternative design approaches, and reviews traffic management and control techniques and their likely impacts. It then sets forth some procedures for evaluating neighborhood traffic management proposals and suggests strategies for implementation.

In moving toward a solution to neighborhood traffic problems, the skills of the politician, lawyer, and budget analyst are often needed. But the solutions tend to be those of planning, design, and operations. This book is especially intended for the creators of such solutions—urban planners, traffic engineers, and the neighborhood residents themselves.

The urban planner will bring to this effort a recognition of the importance of healthy, vital residential neighborhoods to the well-being of the urban environment, as well as an understanding of the social role of the street in the daily life of its residents. The traffic engineer will recognize the role of the street as an artery and as a foot and cycle path, and the relevance of residential street problems to the basic goals of traffic control, which are to enhance:

1. safety
2. efficient movement of vehicles consonant with local conditions and the safety objective
3. environmental goals, such as accessibility for local traffic, minimization of unnecessary traffic, and encouragement of changes in modal choice where this promotes the quality of life and/or specific objectives of a neighborhood

The residents must be part of the process; their contribution is the articulation of values and priorities, their response to proposed plans and designs—perhaps offering ingenious alternatives of their own—and their willingness to assist in the eventual implementation.

Neighborhood traffic management does require ways of thinking that may be new to both planners and engineers. Planners, who have usually focused on land uses, will find a need to pay closer attention to the traffic impacts of their work. Engineers, whose tradition has emphasized maximizing capacity and offering fast, direct routes for the movement of vehicles, will need to give greater consideration to the effects of transportation operations on affected land uses. Both will need to learn to work together to produce coordinated, effective strategies that simultaneously protect residential districts and offer a safe, convenient, efficient circulation system.

Often, opportunities arise to design a new residential neighborhood. Probably even more frequently there may be a call to redesign an old one and to retrofit it with changed geometric street and intersection design elements and with changed traffic controls. Both types of situations are addressed in this book.

The basic problem is not limited to any one country, and this book therefore sets out to treat its subject as an international one. However, problems, procedures, and examples tend to be drawn from the United States and the Netherlands, simply because these are the countries the authors call home. It is recognized that legal systems, political institutions, and economic conditions vary greatly; readers must make due allowance for such diversity and adapt what they read to the constraints in which their society permits them to plan and design. However narrow such constraints may be, there is a challenge to enhance the quality of life on and around residential streets by the use of creativity and imagination. It is hoped that this book will assist the reader in the pursuit of these goals.

SUGGESTIONS FOR FURTHER READING

APPLEYARD, DONALD. *Livable Streets*. Berkeley, CA: University of California Press, 1981.

BUCHANAN, COLIN. *Traffic in Towns*. London: Her Majesty's Stationery Office, 1963. (Also published in a "specially shortened edition" by Penguin, 1964.)

GANS, HERBERT. *People and Plans; Essays on Urban Problems and Solutions*. New York: Basic Books, 1968.

HAGAN, W. B., and B. C. TONKIN & ASSOCIATES. *Physical Design Guidelines for Residential Street Management*. Adelaide, South Australia: 1985.

L. HAWLEY TOWN PLANNING SERVICES and F. R. GENNAOUI & ASSOCIATES. *Evaluating Residential Street Management Schemes: Guidelines and Criteria*. Adelaide, South Australia: Directer-General of Transport, 1984.

JACKSON, KENNETH T. *Crabgrass Frontier: The Suburbanization of America*. New York: Oxford University Press, 1985.

JACOBS, JANE. *The Death and Life of Great American Cities*. New York: Random House, 1961.

LYNCH, KEVIN. *The Image of the City*. Cambridge, MA: The Technology Press and Harvard University Press, 1960.

LYNCH, KEVIN, DONALD APPLEYARD, and JOHN R. MEYER. *The View from the Road*. Cambridge, MA: MIT Press, 1964.

ORGANISATION FOR ECONOMIC CO-ORPERATION AND DEVELOPMENT (OECD). *Traffic Safety in Residential Areas*. Paris: Oct. 1979.

SOUTH AUSTRALIA, DIRECTOR-GENERAL OF TRANSPORT. *Report of the First Seminar on Residential Street Management*. Adelaide, South Australia, Sept. 1983.

URBAN LAND INSTITUTE. *Residential Streets: Objectives, Principles & Design Considerations*. New York: Uban Land Institute, American Society of Civil Engineers, and National Association of Landbuilders, 1974.

2

Residential Neighborhoods and Their Streets

RESIDENTIAL NEIGHBORHOODS

Streets have been called the arteries of city life. Since antiquity, wheeled traffic, drawn by animals or humans, has shared the street space with pedestrians, children playing, adults visiting, itinerant peddlers, residents selling wares in front of their houses, and—where extra space is available—with markets and places for entertainment. There have always been conflicts between the users of the streets. The Romans placed stone blocks at the entrances of some streets as physical barriers against chariot traffic.

In medieval cities transportation functions of streets came into conflict with military reasoning. Cities were contained by fortifications with little room left for expansion. As a result, there was barely enough width for streets and lanes. Facing buildings almost touched each other. Sewage flowed in open channels. Later, changes in military technology caused the leveling of walls and their replacement by wide boulevards giving access to new development areas—the first suburbs (such as Vienna). During times of civil unrest, military reasoning was responsible for leveling entire quarters of cities. Wide thoroughfares were built for quick movement of troops (as in Paris under Baron Hausmann).

After the industrial revolution, streets in expanding cities were dimensioned to let sun and air reach adjacent buildings in order to provide healthy living conditions for workers. In most cities today, streets are used as utility rights-of-way for underground transport of water, sewage, gas, and sometimes steam, electricity, telephone, and cable television lines.

4

Only in the late nineteenth century did streets become the domain of vehicles. Historically, sidewalks were used to protect the pedestrian from mud and debris. As the number and use of vehicles grew, however, sidewalks became a safety feature. First horse-drawn, then motor-powered vehicles constrained pedestrian movement to sidewalks and designated crosswalks. Once-shared street space became divided by functions.

The widespread use of the automobile in the twentieth century has further increased vehicular dominance of street space. As motor vehicles became more pervasive, planners and engineers directed attention and energy toward facilitating their safe and convenient operation. Roadway design standards were developed with the motor vehicle in mind, often with the effect, explicitly or implicitly, of ignoring or banishing many of the historic users.

While the mobility provided by vehicles has always been valued, it also has always been apparent that it came at a cost. Since the latter half of the nineteenth century, concern about traffic management in neighborhoods has been part of the western planning experience. Much of the early thinking was focused on questions of design for new communities which would by their physical structure and amenities provide a superior quality of life for their residents. This early thinking soon came to be reflected (in part) in designs for residential subdivisions, with curvilinear streets, cul-de-sacs, landscaping, and open space used to restrain traffic and create a peaceful, sometimes even bucolic, setting. Only more recently, however, has much attention been given to traffic management in established neighborhoods, or has recognition been accorded to nonmotorized street users and abutting residents.

This chapter explores the history of neighborhood planning and related research in environmental design and in the social sciences, and points out their influence on transportation planning for neighborhoods. The assessment of these planning concepts leads into a discussion of the issues concerning traffic in neighborhoods today, with a focus on the needs and values of street users and residents. The chapter then states planning goals for design and management of traffic on residential streets, and proposes a policy framework that treats streets as "public places," not dominated by one user group alone. The chapter concludes with a brief review of commonly used definitions of street functions and design standards, and a discussion of traffic flow on residential streets.

Neighborhood Planning and Street Design: A Brief History

In general, there have been two design themes for residential neighborhoods, the "gridiron" and the curvilinear street pattern.

The gridiron rectangular network, already found in cities of antiquity, has been commonly used in many countries because of the simplicity in surveying new residential areas and writing descriptions of individual parcels of property (Reps, 1965). Prior to the twentieth century, a system of street hierarchies was considered relatively unimportant; note as an example, however, that in the strict gridiron pattern on Manhattan Island north of 14th Street, some east-west streets were allocated wider rights of way because of their envisioned status as major thoroughfares. The gridiron pattern used a variety of dimensions in different locations, from the very short street spacings of Manhattan to the very generous ones of more recent cities. With the rapid growth of urban concentrations in the late nineteenth century, the gridiron plan became the preeminent form given to American cities.

One disadvantage of a rigid adherence to rectangular street layout was that topography often interfered with such simple schemes. Surveyors, who ignored the existence of hills in their unswerving faith in gridirons, sometimes laid out streets which could never be constructed, such as in San Francisco, where the pattern even extended under water into the Bay, or in Philadelphia, where the grid was intended to cross steep bluffs and ravines (Fig. 2.1).

As a reaction to the ubiquitous rigidity of the gridiron system, some communities took steps to break up the street pattern into neighborhood units. For example, closing off city streets to form private neighborhoods, sometimes using elaborate wrought iron gates

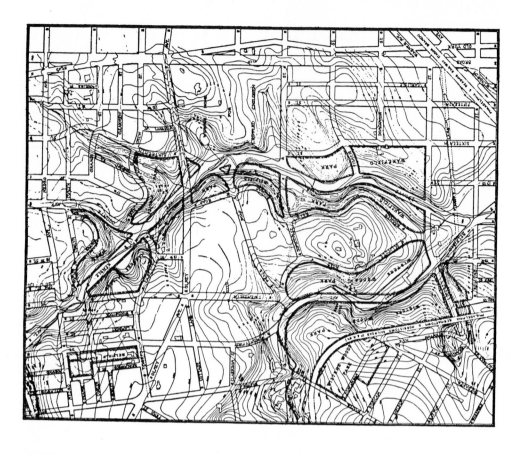

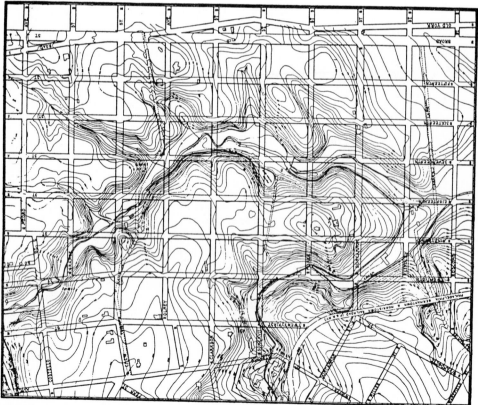

Figure 2.1 Part of original street plan for Philadelphia, and the same section as actually built.

and brick pillars, began in St. Louis in the 1860s. The curvilinear subdivisions designed by Frderick Law Olmstead and Calvert Vaux at the end of the last century are also part of the countertradition to gridiron streets (Fig. 2.2). However, these planned or re-planned street networks were the exception. In the vast majority of areas, the street systems were simply laid out along the surveyed frontage lines by real estate developers.

It was in the 1920s that major adverse impacts of automobile traffic in typical gridiron streets began to be felt. Planners, to alleviate ever worsening conditions, now began to define a hierarchical street pattern and differentiated between high capacity streets and local, frequently curvilinear, residential streets. In an attempt to define the neighborhood of the future, a concept was developed that had its roots in Ebenezer Howard's "Garden City": a planned neighborhood that combined the good qualities of urban life with the good qualities of rural living, but eliminated the pathological effects and dreary uniformity of crowded nineteenth century mass inner-city housing (such as in Britain) as well as un-sanitary rural living conditions frequently isolated from educational and cultural facilities.

Clarence Perry gave definition to this concept in North America. While living at Forest Hills Gardens, Perry conceived of an "investment for social betterment" and a model of community development which, through the coordination of highways, open spaces, and common facilities, could benefit the social aspects of neighborhood life. From direct observations of life in this community, Perry deduced and analyzed factors he felt were responsible for its success and then proceeded to reduce them to a set of general principles

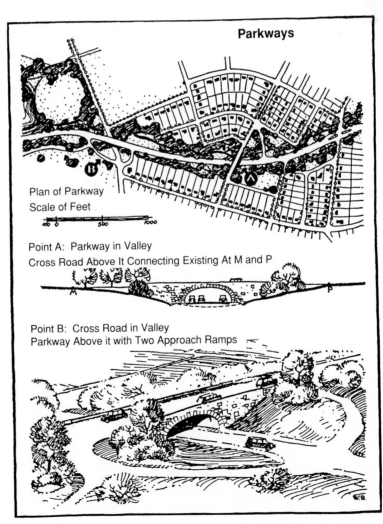

Figure 2.2   Subdivision with curvilinear streets. (Olmstead and Vaux)

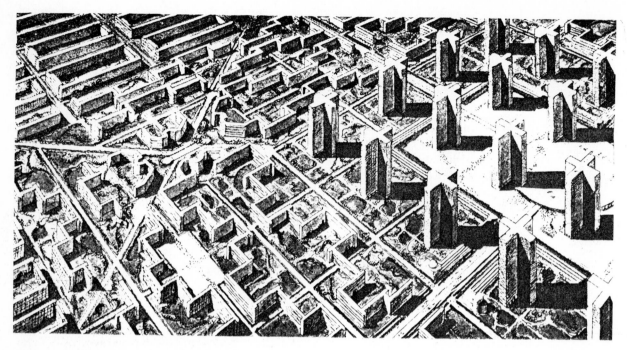

Figure 2.3   Cité Industrielle. (Le Corbusier)

so that they might be widely emulated (Perry, 1929). Indeed they were. The most influential planners and architects of his time used these principles, including Lewis Mumford, Le Corbusier (Cité Industrielle, Fig. 2.3), Frank Lloyd Wright (Broadacre City, Fig. 2.4), and Clarence Stein (Radburn, Fig. 2.5).

The concept was relatively easy: a neighborhood unit was defined with definite boundaries—in most cases major streets with heavy traffic (Fig. 2.6). Within these boundaries internal streets gave access to an area, housing a population of predetermined size, "not too large to destroy personal contact and not too small to fail to afford variety and diversity" (Pick, 1941). Five thousand persons was seen as an optimum, largely because a population of that size could then support a primary school. This school in the center of the community, adjacent to a public open space, was in easy walking distance from all parts of the unit. The walking distance together with population defined the extent of the unit and its density. Like the wagon trail camps of the early settlers, the emphasis was on safety and identity through containment. Safety was to be further promoted through a network of pedestrian paths linking all parts of the unit with the center and the commercial facilities, if for no other reason than that "children should never be required to cross a main traffic street on the way to school" (Figs. 2.7, 2.8). This physical plan was viewed as providing spatial order in the chaos of surrounding environments, both urban and natural. It also was seen as having high social and psychological significance: the unit provided "a clear identity in people's consciousness." The emphasis was on "face-to-face contacts," prevent-

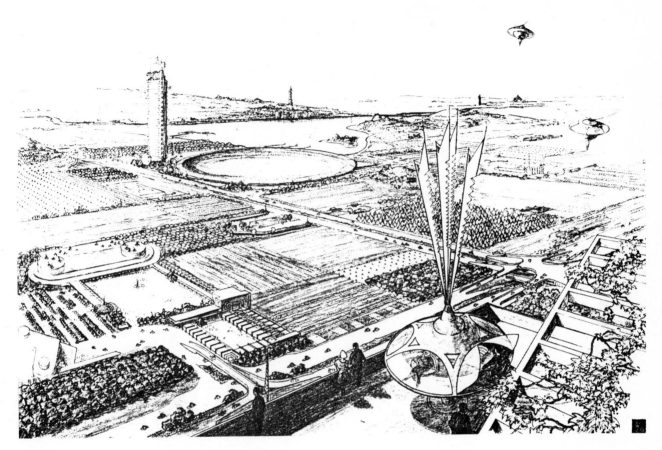

Figure 2.4 Broadacres. (Frank Lloyd Wright)

ing the feared rootlessness and anonymity of the cities and fostering loyalty, security, and a feeling of belonging.

While much of this planning philosophy was directed to bringing about identity through shared experiences inside the neighborhood, it was somewhat weaker in the emphasis on diversity and tolerance. A mix of income groups and housing types that would bring about diversity was one of the goals, and it was proposed that, through the "conscious practice of democracy in small units" (Mumford, 1945), social barriers would break down and provide a meaningful role for the individual in mass urban society. But it was at the same time argued that "sameness" would foster community spirit at the local level. The combination of social idealism and physical planning inspired planners to advocate the concept of a standard unit which would "provide a panacea for all problems of residential development and somehow fit all cases and needs" (Pearson, 1939).

Figure 2.5   Radburn. (Clarence Stein)

The Radburn plan (1929), designed by Henry Wright and Clarence Stein, was a true blueprint of Perry's concept (Fig. 2.5). It was radically new in its separation of car traffic from pedestrian traffic. Overpasses and bridges enabled pedestrians to walk from one end to the other without ever encountering traffic. Although the concern for safety, both implicit and explicit, found in the Radburn Plan and in Perry's schemas did not take hold immediately, the growing volume of traffic on the highways of the United States increasingly drew federal attention during the 1930s and 1940s. The *Community Builders Handbook*, published by the Federal Housing Administration, advocated a hierarchical system of major, collector, and minor access streets that became a standard throughout urban America. But the disadvantages of gridiron networks continued to exist and even increased. Every street offered a possible route through a neighborhood and therefore was likely to attract through traffic; traffic engineering measures to enhance movement on those streets designated as "major" served to minimize through traffic on purely local streets, but only until the major streets became overcrowded. And, of course, many of these major streets received such a classification long after adjacent land use had been established and often where such land use was incompatible with fast through traffic.

Thus it was that in the late 1940s and early 1950s, Montclair in New Jersey and Grand Rapids in Michigan became the first cities to install traffic diverters and convert neighborhood access streets into cul-de-sacs as a reaction to this state of affairs.

With the introduction of large-scale urban redevelopment projects in the 1950s and 1960s, a renewed attempt was made to reduce traffic on neighborhood streets and offset the social and fiscal costs of gridiron planning. Street closures and the wholesale clearance

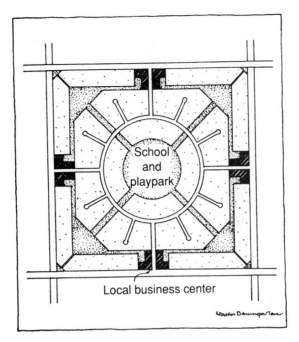

Local business center

Figure 2.6 Theoretical pattern of a neighborhood.

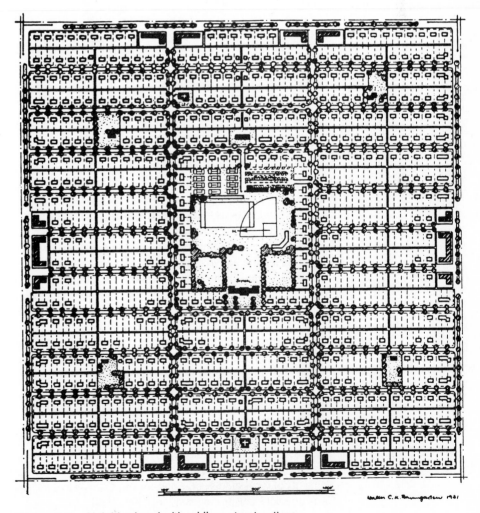

Figure 2.7 Neighborhood with gridiron street pattern.

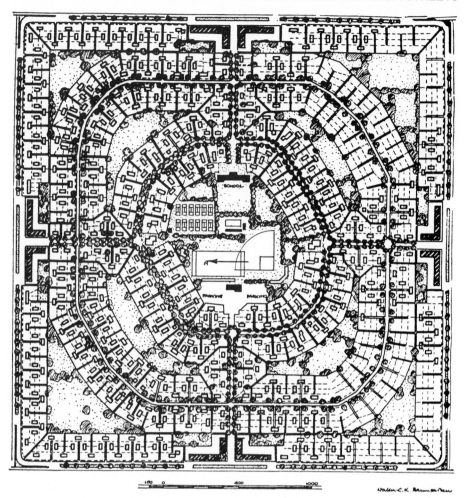

Figure 2.8   Neighborhood with curvilinear street pattern.

of residential sections to create new superblocks occurred in cities as distant as Baltimore, Los Angeles, New York, and San Francisco. The design of superblock housing projects often limited automobile access and created pedestrian cores at their center.

As an alternative to the superblock, some large, high-density residential developments in central areas have followed the Radburn principle of vertical separation of pedestrians and the automobile. In such developments, the pedestrian activity is moved to a level above that of the street network. Various blocks of the development are connected by pedestrian bridges, so that it is possible to walk from one end of the neighborhood to the other without having to cross any street at grade. Residential buildings have parking garages at and below the street level, and "front doors" are located on the pedestrian level above the garage. If retail uses are included in such developments, they are usually placed at the street level to facilitate delivery of merchandise. Frequent stairs and, perhaps, escalators, connect the pedestrian podium with the street level. But this style of development is still uncommon.

The construction of networks of urban freeways after World War II had an unintended, significantly deleterious side effect on city and neighborhood traffic patterns in many U. S. cities. Freeway ramps were planned with emphasis on the traffic-carrying capacity of local streets or the possibility of converting pairs of such streets into major one-way couplets, penetrating many neighborhoods. In some cases, the environmental quality of neighborhoods was compromised by the increase of freeway-generated traffic on streets

designed for purely local access. Sometimes streets were widened, requiring removal of
rows of residences and consequently resulting in major damage to or even destruction of
the neighborhoods. Other "improvements" in the form of increased illumination, coordinated signal systems permitting speeds of well over 25 mph (40 km/h), and conversion
to one-way flow impacted negatively on residential streets.

Disenchantment with gridiron networks and the desire for containment resulted in
irregular street patterns (Fig. 2.9) in new subdivisions, with only one or a few points of
connection to the major street system. A new problem appeared, however, as each subdivision developed a plan unrelated to those of its neighbors. With entry and exit points
often located only along one side of the property being subdivided, sequential trips were
subjected to extensive circuity—in some cases causing bus operators to refuse to establish
routes within the neighborhood.

The first significant planning guide to specifically address the traffic problem in
residential streets was published in Britain in 1963 by Colin Buchanan. Britain had adopted
the neighborhood unit principle with the Dudley Report (Design of Dwellings, 1944). The
rapidly growing housing demand after World War II was met by New Towns or satellite
towns ringing old metropoli. The old towns, and especially the inner-city neighborhoods,
suffered greatly from the influx of commuters on their way to work from the suburbs into
the inner cities. In his report to the Ministry of Transport, "Traffic in Towns," Buchanan
(1963) wrote:

> The briefest acquaintance with the conditions that now prevail in towns makes it
> clear that traffic congestion has already placed in jeopardy the well being of many
> of the inhabitants and the efficiency of many of the activities. Unless something is
> done about the potential increase in the number of vehicles that come into neighbor-

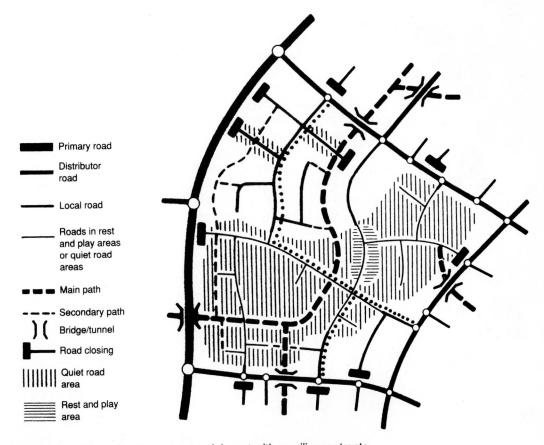

Figure 2.9   Hierarchical street network layout with curvilinear streets.

hoods the conditions are bound to become extremely serious within a comparatively short period of years. Either the utility of vehicles in towns will decline rapidly,
or the pleasantness and safety of surroundings will deteriorate catastrophically—in
all probability both will happen together.

Buchanan advocated measuring the capacity of streets in neighborhood in a different
way from that used by the traffic specialists. He postulated that neighborhoods had an "environmental capacity" which placed limits on the amount of traffic they could absorb. He
defined "Environmental Areas" from which extraneous traffic should be removed and the
system reorganized for internal movements by vehicles and pedestrians. The traffic
capacity would be kept low due to "limited width or inadequate junctions." That would
serve as a general restraining valve, and traffic would be diverted around such environmental areas. Buchanan argued that the capacity of a neighborhood to absorb traffic is a
function of measuring a set of environmental qualities that take into account acceptable
levels of air pollution, noise, visual disfigurement, luminosity, safety, and pedestrian activity.

Buchanan admitted to the broadness of descriptions of environmental qualities and
the absence of detailed standards and effective units of measurement. His method necessarily lacked precision but it pointed in the right direction: the planner should be concerned
with resident satisfaction.

In the same period, however, challenges to physical planning and especially to the
physical planners' conceptions of social betterment through physical design were beginning to surface, particularly in the United States. Social scientists' research findings ques-
tioned whether physical layout principles had a major effect on resident behavior patterns
and social interaction. Researchers admitted to some evidence supporting the important
effect of the physical layout on functional and social interactions. However, their findings
increasingly recognized other factors, such as the characteristics of the residents, degree
of mobility, social values, norms, attitudes, and other determinants as the dominant influence on social behavior (Gans, 1967).

In addition, the emphasis of planners on order and aesthetics were blamed for such
policies as massive urban renewal programs that cleared away low-income neighborhoods
to make way for high-rise apartments for middle and high-income residents, and for commercial and civic facilities; closing of streets to form superblocks was a common con-
comitant of many of these schemes.

Others argued that traditional concepts of neighborhood planning were outdated, because people's sense of community had radically changed. Webber, for example, pointed
out that widespread ownership of the motor vehicle had enabled individuals to travel at
high speed and low cost to any part of the metropolis. The ubiquitous availability of
telephones had further supported wide-ranging personal contacts. The result, he found, was
a movement away from a local society in which roots, place, and propinquity were the
determinants of social relationships, toward a society in which there was an increasing
reliance on friends rather than neighbors. (Webber, 1964.) He and others argued that the
physical planners' desire to order urban society was not only inconsistent with social behavior, but was "undemocratic"; they viewed mobility, diversity, and change as critical
elements in personal freedom.

Indeed, the 1950s and 1960s had brought an increasingly physical and social
mobility. In many communities, neighborhood planning waned; emphasis was placed on
maintaining and enhancing freedom of movement. Transportation planning, in particular,
placed emphasis on mobility. Standards for local streets were developed which called for
12-foot travel lanes in both directions with parking lanes along both curbs. Major streets
were often twice as wide.

While the social scientists had attacked the kind of physical planning that was overly design-oriented with untested assumptions on social issues, laying too much emphasis

on physical order, their vision of a placeless, mobile community was limited in its own way. The costs of mobility and of the de-emphasis of neighborhood concerns soon became apparent. The poor were the first to rebel against the society of "unlimited choices"; neighborhood actions during the late 1960s and 1970s demonstrated the strength of local groups to speak against planning proposals that would continue to uproot people. Unrest soon spread into middle-class neighborhoods as concerns for the natural environment grew into a broad movement emphasizing nongrowth, or growth management of metropolitan areas. The energy crisis in the 1970s further fueled the discussion of placing limits on the consumption of resources. "Society has become more placeless, but the phenomenon of wanting that which we no longer possess extends to the new values of the natural and visual environment." (Appleyard, 1981b).

With increasing concern about environmental values, Buchanan's concept of carrying capacity and resident satisfaction received renewed attention. Moreover, the question of the *dimension* of a neighborhood became secondary; Hester's definition of a neighborhood as a political settlement comes closer to the heart of the matter (Hester, 1975). In this view, it is through shared concerns and problems that local loyalties or roots are strengthened and common values develop. Depending on the issue and extent of the problem, the dimensions of the neighborhood will vary. Boundaries will depend on social class, life cycle, lifestyle, ethnicity, and personal preference. A problem can affect several thousand people or a few hundred, or just a dozen families on one street. The "neighborhood" in this sense is not the physical community, but the community of interests.

Although the boundaries of what is the "neighborhood" may not be fixed, people are very likely to be concerned about the areas close to home, and about helping to create healthy environments that satisfy residents' needs and expectations. Because local streets and their traffic are important elements of the urban environment, responsible neighborhood planners and traffic engineers must consider these matters and reflect these community concerns in their work.

Neighborhood Values Concerning Streets

What are the things people value in their neighborhood? A study by Appleyard and Lintell (1972) of the environmental quality of city streets in San Francisco showed a strong relationship between traffic volumes and such values as safety, security, identity, comfort, neighboring, privacy, and home. Three streets selected for this analysis were identical in appearance but quite different in the traffic volumes they supported. One street had "light" traffic, some 2,000 vehicles per day; another "medium" traffic, 8,200 per day; and the third "heavy" traffic volumes, 15,750 vehicles per day. This latter street had a one-way traffic flow with synchronized signals that encouraged speeds of up to 45 mph (72 km/h); the two other streets were both two-way with lower average speeds. All the streets formed part of the same residential Italian neighborhood, with a mix of other groups of European-Americans and a small but growing Oriental population. By social class and income, these streets were relatively homogeneous. Differences occurred in length of residence, family composition, and ownership patterns.

The results of this study, based on extensive interviews and field observations, revealed that residents on all streets were primarily concerned with the dangers of traffic. Excessive speed on the "heavy" street was frequently cited as dangerous, although most often this danger was experienced indirectly by witnessing large numbers of speeding cars driving downhill in one direction, or by hearing the high-pitched squeal of brakes. Residents on the "light" street were also preoccupied with safety; the random appearance of a "hot rodder" without warning was considered a greater menace for children and older people than the steady stream of traffic experienced on the "heavy" street. Other factors taken into account in the study included measurements of noise, vibration, fumes, soot, and trash. In each case an inverse correlation existed between the adverse environmental impact measured and the degree of livability experienced. For example, on the "heavy" street, the least desirable overall in livability, noise levels were above 65 decibels for 45 percent of the time measured, and did not fall below 55 decibels for more than 10 percent of the time. The "medium" street had noise levels of 65 decibels or more 25 percent of the time, and the "light" street experienced a 65 decibel limit only 5 percent of the time (Appleyard, 1981a).

When residents in the case study were asked whom they knew and talked to on their street and were asked to identify where people met socially, an inverse correlation was evident between the volume of traffic and the amount of social interaction (Fig. 2.10). Similarly, inverse correlations were found between traffic volumes and two contrasting factors: feelings of neighborliness on the one hand, and perception of privacy on the other. In a similar way, the residents' sense of identity with the environment was more curtailed the greater amount of traffic they had to deal with. Intensive traffic conditions on the "heavy" street, and to a lesser degree on the "medium" street, led to personal stress and suffering. Social disruption on the heavily trafficked street resulted in a greater rate of occupancy turnover. This was especially true for families with small children.

A follow-up study of former residents of the "heavy" street found that the reason they gave for leaving centered on traffic impacts and crime levels. Many of these former residents were obliged to leave San Francisco once they decided to move, forced out of the city because of the shortage of suitable accommodations. Those that stayed on adapted (or perhaps became resigned) to their environment. Since many of these people were older and living on retirement incomes, adaptation was their only option, with no future change for the better in sight. Evidence of adaptation to environmentally unsatisfactory conditions found by Appleyard substantiated that of other researchers.

Many of the findings of Appleyard's study seem obvious, especially the inverse correlation between traffic flows and levels of livability. However, the implications of these findings were far-reaching. They changed official city policy in San Francisco. The *San Francisco Urban Design Plan* (San Francisco City Planning Department, 1971) proposed "protected residential areas" throughout the city. These areas were to be protected from through traffic by policies such as

- the improvement of public transit
- the concentration of traffic on major streets by increasing the capacity of these facilities, and
- discouraging through traffic on minor streets by treatments such as rough pavement surfaces, widening sidewalks (especially at intersections in the form of "chokers" as described in Chapter 5), curved alignments, landscaping, and lighting, all designed to slow traffic down to a residential pace

In instances where traffic flow could not be reduced, tree plantings as a way of ameliorating conditions were proposed. However, these proposals were not received without critical comment.

The principal change in the plan would involve the creation of hundreds of barriers to impede the flows of through-auto movement, a reversal of previous city concern and effort to speed traffic up. (*San Francisco Chronicle*, May 3, 1971)

This juxtaposition of thinking describes exactly the anxiety with which transportation planners and public works directors received Appleyard's study. In the past, the objective had been to facilitate traffic; the objective now was to manage it.

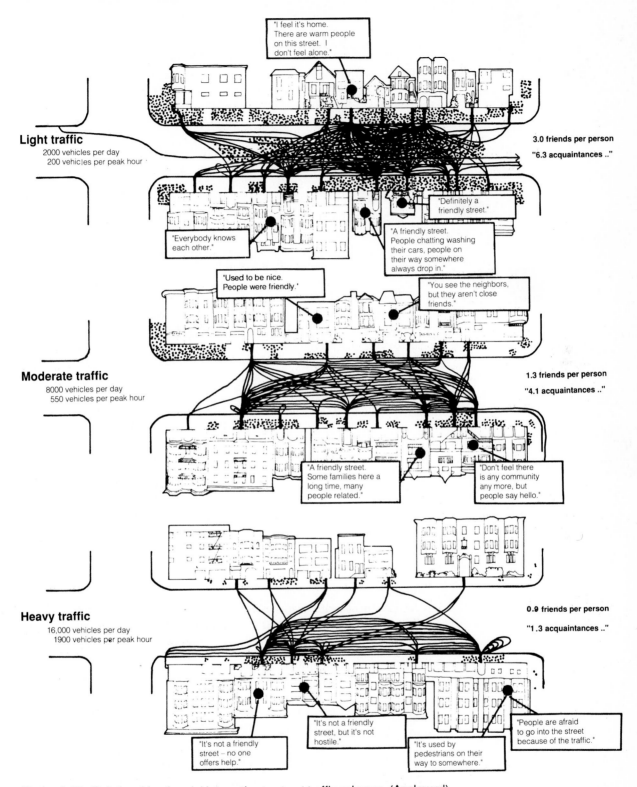

Figure 2.10 Relationship of social interaction to street traffic volumes. (Appleyard)

Safety Concerns

Appleyard's study documented the perceptions residents have of traffic hazards and exposure to possible accidents. There are few studies that try to quantify the real hazards, since neighborhood or street types are not always reported in accident data and specific studies are scarce. Furthermore, traffic accident statistics, even if they were available, would not define the full extent of the concern over street safety.

> It is precisely in the residential environment that there is more to road hazards than is evident from traffic accidents: the feeling of being unsafe, the experience of a certain threat emanating from traffic. (OECD, 1979, p. 4)

Where such feelings are prevalent, the elderly no longer gather on the street for a chat or to go for a walk; children are not allowed to play in the front of their homes and are brought to school by their parents because the walk to school is considered too dangerous. These are expressions of a perception of hazard (Untermann, 1984).

It is difficult to analyze accidents in residential areas scientifically. Most underlying data do not clearly distinguish the type of area in which accidents were reported, and most published reports reflect this data deficiency. It is also difficult to define an exposure measure to be used as a base for calculating trends or making comparisons; the vehicle-mile unit has obvious shortcomings when pedestrian accidents are being studied.

However, against the background of continuing concerns about safety in residential areas, it may be useful to provide a sample of experiences reported in the traffic safety literature:

— Australian researchers report that between one quarter and one third of all *reported* urban casualty accidents were occurring on neighborhood streets in the early 1980s (Brindle, 1983). In Melbourne, 38 percent of reported bicycle accidents occurred on roads ranked below Second Arterial Class (Brindle and Andreassend, 1984).

— In Germany, it was found that the size or density of population in a neighborhood has no systematic effect on accident risk, but that new areas are safer than older ones. A major factor contributing to accidents was the presence of roads cutting through the neighborhood as contrasted to peripheral roads (OECD, 1979, pp. 86-80).

— Studies from the Netherlands (Verkeer en Waterstaat, 1982; SWOV, 1980) include the following findings:

  • Accident occurrence is spread over the entire urban area, and only seldom is there evidence of a concentration of accidents on residential streets.

  • The majority of accidents in residential neighborhoods occur on collector-type streets.

  • A nondifferentiated street network is less safe than a hierarchically arranged pattern.

  • Hazards are greatest on streets where there is much parking, on streets with many intersections, on long, straight streets, on busy streets, and in districts with few playground facilities.

  • Pedestrians often have accidents when crossing roads from between parked vehicles; in particular, parked cars block the view of the road for children and the view of children for approaching drivers.

  • The proportion of children and elderly persons involved in accidents is greater on residential streets than on other streets.

  • There seem to be no serious accidents involving injury at driving speeds of 30 km/hr or less.

- Road safety is greater in newly built residential districts than in older neighborhoods.

The safety of children is of particular concern. It is difficult to prevent children from occasionally playing in the street or, more frequently, from running into the roadway in the progress of play; it may, in fact, be desirable to have children ride bicycles in the roadway as a learning experience for the time when they will use bicycles as a general transportation mode. The safety literature on traffic accidents involving children includes the following findings:

— Most accidents involving children occur in the vicinity of their homes, as might be expected. One report (Bennett, 1974) reports that 84 percent of children aged less than 10 in four London boroughs were injured within 800 meters of home. Similar findings were reported in Verkeer and Waterstaat 1982, SWOV 1980, and Wade et al. 1982.

— In Germany, 55 percent of accidents involving children happened on roads in residential areas (Pfundt, 1977).

— Seventy percent of all accidents in the Netherlands involving children under 6 occur on streets carrying fewer than 3,000 cars per day (Bakker 1974).

— In a study of over 2,000 pedestrian accidents in 13 U.S. cities (Snyder and Knoblauch, 1971), slightly over 50 percent of all accidents involved youngsters under 15 years of age.

— In Great Britain, a 1978 report showed that pedestrian deaths for children aged between 5 and 9 represented 18.6 percent of all traffic deaths in this age group, compared to 5 percent for the total population; for the 10- to 14- year age group the percentage was 11.4 (Foot et al., 1982).

— Accidents involving children up to 5 years old often occur between intersections on local roads, while older children are more often involved in intersection accidents and on busy streets (SWOV, 1980, Verkeer en Waterstaat, 1982).

From safety research, with special emphasis on traffic safety of children, the Organisation for Economic Co-operation and Development (OECD) drew the following conclusions (OECD, 1983):

— Strict differentiation of streets according to their function leads to safe residential areas;

— Full separation of vehicle, pedestrian, and cycle movement is accompanied by very low accident rates;

— Cul-de-sac streets are safer than loop streets and considerably safer than ordinary through streets;

— On those roads providing a distributive function, accident rates are minimized where frontage access is prohibited and the layout of the residential development is such that pedestrians and cyclists have no need to use routes that run alongside highways.

— Space-sharing techniques like "Woonerven" (see Chapter 4) make drivers more aware of their responsibilities toward the more vulnerable groups.

Conflicts

The concern for safety on residential streets is a guiding factor in street design. The problem of safety, however, is rooted in the conflicts that exist among users of street space. Streets are public property, and therefore belong to everyone. In reality, however, some users have preempted more of the street space than others, as mentioned at the outset of this chapter. In specific traffic engineering terms, the traffic function—moving traffic streams efficiently—competes with the land service function of providing access to properties, parking in the street, and so forth. Fig. 2.11 illustrates these competing demands and how they are often arbitrated by designation of road types.

In order to identify goals for residential street design, it is important to understand who the groups are that participate in streets and to understand the full range of opportunities streets have to offer. Appleyard refers to five principal conflicts over the use of street space:

1. conflict between travelers and neighbors
2. conflict among the travelers themselves, especially between drivers of cars, trucks and buses, pedestrians, cyclists, and so forth
3. conflict between the neighbors, especially residents and neighboring merchants, institutions, or industries
4. conflict between the public agencies that manage and maintain streets and protect neighborhoods (such as public works, police, fire services) and the neighbors
5. conflicts among the professionals who plan, design, and manage streets, chiefly between engineers and designers

On residential streets, then, the conflict between traffic and livability is pervasive. Traffic may threaten even when infrequent or out of sight, if the occasional vehicle passes through at excessive speed. The conflict is most poignant when it involves pedestrians (especially children) and cyclists. It is often most controversial when it pits residents against adjacent institutions, such as hospitals, offices, schools, merchants, and even transit agencies operating large bus or rail stations, which generate traffic and spillover parking demand. The sanctuary of the neighborhood is at odds with the operation of the traffic artery, the resident with the traveler. Of course, residents are at times travelers themselves, and thus are in conflict with their neighbors.

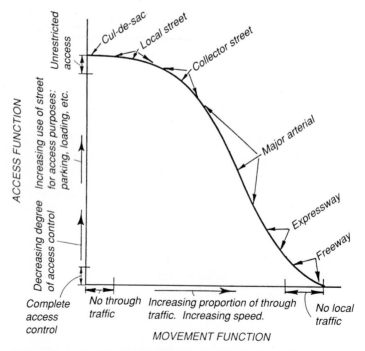

Figure 2.11 Schematic relationship between access and movement function of streets

Until recently, and in many cities today, public works departments have tried to satisfy the demands of the travelers. Streets have been widened at the expense of sidewalk width, rows of trees, and sometimes front yards. City agencies have made these changes based on the assumption that traffic would increase at a steady rate into the distant future. The principal measures of street use were traffic volumes and capacities. Bicycle and pedestrian traffic was seldom estimated or considered in the design.

Fire and police departments argue for easy access. Public safety in emergencies is a dominant public welfare concern, even though emergencies are relatively rare. Many emergency services could be more flexible and vehicles smaller in size; for example, small hook-and-ladder trucks have been developed in the Netherlands. Similar considerations apply to garbage collection and moving vans. However, such possibilities are seldom given attention, in part because responsibility for planning and delivering these services are scattered among several public and private agencies.

Goals

This list of conflicts suggests the importance of developing a sound public policy for designing and managing streets with all users in mind. In many cities it has become increasingly the policy of public agencies to support the weaker users, such as the very young and old, pedestrians, the handicapped, and those without access to automobiles.

In solving the conflicts that exist on residential streets, it is essential that designers develop compromises and consider tradeoffs. Professionals with creative ingenuity can design streets to serve several groups' needs within limited space. They can plan to use each street to its "environmental capacity," to imagine new functions and meaning for the street, and to return the street to the center of neighborhood life.

DESIGN OF LOCAL STREETS AND TRAFFIC CHARACTERISTICS

To begin to address the conflicts that exist on streets in residential areas, it is useful to review current practice in street classification and design. These current practices, it will be seen, strongly affect the patterns and types of traffic that are found on these streets.

Streets as Traffic Facilities

Transportation Functions of Streets. Streets (and rural roads, for that matter) evolved over many centuries to perform two transportation functions: provision of *access* to individual parcels of land, and provision of an infrastructure for *movement* between various origins and destinations.[1] However, these functions make competing demands on the street; in most situations a tradeoff must be made as to the relative importance of access and movement.

"Access" can be interpreted to include the existence of driveways connecting the street with private property and the availability of part of the street for parking and loading.

"Movement" comprises both the *capacity* to move quantities of vehicles or people and the ability to do so at a reasonably high *speed*.

Hierarchy of Streets. In planning street networks to serve these functions, three categories of streets have been developed: one emphasizes the local access function, one optimizes the movement function, and one is an intermediate compromise.

---

[1] They also serve to provide many nontransportation purposes, as discussed at the beginning of this chapter.

1. *Local streets* provide access to immediately adjacent properties. Through movement may be possible, but is not encouraged by operational controls; it may be impossible in the case of cul-de-sacs. Part of the street width is usually allocated to vehicle parking without restrictions, although special snow emergency parking prohibitions may be necessary. Each abutting property may have a driveway connection to the street.

2. *Collector streets*, while providing access to abutting land parcels, also enable moderate quantities of traffic to move expeditiously between local streets and the major street network.

3. *Major streets* are designed for the efficient movement of through traffic at speeds which are as high as can be reasonably allowed in view of safety considerations and the amount of access also being provided. Capacity is obtained by provision of wide street cross-sections and high-capacity controls at intersections, or by elimination of intersections by grade separation. Speed results from provision of good horizontal and vertical alignments and removal of potential safety hazards, especially access friction.[2]

Major streets continue to perform some access functions, but these may be limited by parking restrictions to make the curb lane available for through movement in peak periods. In new subdivisions, individual lots may be laid out in such a fashion that none front on the major streets; this obviates the need for driveways and parking on the major streets.

Design Standards

Agencies in many nations have developed design standards for streets. Such standards emphasize safe and efficient vehicle operation, including:

— the maneuvering characteristics and dimensions of the largest motor vehicle likely to use a street
— efficient levels of operating speeds (for major and collector streets)
— assumptions about the extent of on-street parking
— accepted needs to enhance safety through nighttime illumination, prompt rainwater removal, and separation of pedestrians and bicyclists by provision of sidewalks and bicycle paths.

A typical set of design standards recognizes the development density of the neighborhood and the general terrain (level, rolling, or hilly) as well as the functional type of street. For example, one set of standards for residential developments has been recommended by the Institute of Transportation Engineers (1984). Table 2.1 lists some of these standards; there are also policies regarding the provision of alleys, off-street parking, street lighting, and driveway design. The reference should be consulted for the complete set of standards.

In the case of the design standards shown in this table, some concern has been expressed about the upper ranges of pavement width and curve radii for local streets. Generous curve radii enable speeds to be higher than might be desirable. Ample pavement width, when not used for curb parking, also encourages higher speeds; when the street is located near a major traffic attraction, the extra width may be used by overflow parking.

Some policies advise that traffic safety will be enhanced by the selection of building setbacks which will permit parking of vehicles in driveways without encroaching on sidewalks, and by spacing driveways to permit efficient use of the curb for parking. In some areas (such as parts of Evanston, IL, Beverly Hills, CA, and Vancouver, B.C.) alleys

---

[2] In the case of freeways, access is eliminated completely (legally by purchasing access rights from abutting land owners and physically by fencing).

TABLE 2.1

Typical Design Standards for Streets in Residential Areas

Design Item	Local Streets		Collector Streets	
	ft	m	ft	m
Right-of-way width	50-60[a]	15-18[a]	70	21
Pavement width	22-36[a,b]	7-11[a,b]	36-40[a]	11-12[a]
Type of curb	Vertical*		Vertical	
Sidewalk width	0[†]-6[a]	0[†]-2[a]	4-6[a]	1.3-2[a]
Separation between curb and sidewalk[†]	6	2	10	3
Minimum sight distance	110-200[b]	35-60[b]	150-250[b]	45-75[b]
Maximum grade	4-15%		4-12%	
Minimum centerline radius of curves:				
Superelevated	110-250[b]	35-75[b]	175-350[b]	53-105[b]
Not superelevated	180-430[b]	55-130[b]	280-580[b]	85-175[b]
Minimum tangent between adjacent curves	50	15	100	30
Maximum cul-de-sac length	700-1000[c]	210-300[c]	—	—
Minimum radius of right-of-way for cul-de-sac turnaround area	50	15	—	—

[a] - Varies with density of development.

[b] - Varies with terrain.

[c] - Varies inversely with density of development.

* - Rolling or no curb in low-density developments on level terrain.

[†] - No sidewalk in low-density developments.

are provided at the rear of all residential lots for vehicular access to garages or other parking spaces, and the street itself is not encumbered with driveways and the possible hazards related to them. Such alleys may also serve as utility rights-of-way.

These design standards are silent, however, on the relationship of street use to abutting land development and its users; it generally is assumed that other regulations (such as zoning and building codes) will address these concerns.

While the typical street design standards undoubtedly help to assure motorist safety and convenience, they may also inadvertently contribute to certain kinds of problems. For example, wide rights-of-way, long sight distances, and large-radius curves facilitate driving at speeds well above the 25 mph (40 km/h) generally considered a suitable speed limit on local streets. In some cases, excessively wide streets may be an aesthetic problem, creating a barren interruption of neighborhood structure, and may present a formidable barrier to be crossed, especially by the elderly and children. Traffic controls can also contribute to speeding problems by assigning preferential right of way at several consecutive local street intersections to the same street, giving it the perhaps unintended status of a collector or major street.

Many neighborhoods predate current standards, and their streets may have other design problems. For example, it is not uncommon for older major streets to be lined with houses, either with numerous driveways presenting potential conflicts or, in some older areas, with no driveways and no provision of off-street parking at all.

Another concern is that design standards usually are applied at the level of a subdivision or, at most, a local jurisdiction, at least in the United States. This reflects the fact that local street networks are the responsibility and prerogative of municipal governments. Regional transportation planning agencies attempt with varying amounts of success to ensure that a cohesive network of major highways exists and is operated effectively—that routes for through traffic, bus routes, and service and emergency vehicles are reasonably devoid of circuity. However, there are many cases in which adjacent jurisdictions fail to coordinate their networks and streets change designation and character as they cross jurisdictional boundaries.

Traffic on Local Streets

Traffic on local streets often reflects the design characteristics of the streets themselves, as well as the land uses that they serve and the uses in adjoining areas. In order to develop a clear picture of the traffic problems on existing neighborhood streets, and to avoid the development of such problems in new residential developments, it is important to understand the types of traffic which use local streets, as well as the quantities of traffic that can be expected. These matters are discussed below.

Types of Traffic. Traffic on local streets is composed of two types: *through* traffic, which is making trips with both trip ends outside the neighborhood under consideration, and *local* traffic, whose travel patterns include one trip end within the neighborhood. No generalizations can be made about through traffic: its composition varies with the type of urban area, and with the geographic position of the neighborhood within that area.
Local traffic can be divided into six categories; three of these relate directly to the land use of the neighborhood, two to miscellaneous service traffic, and one to possible spillover from adjacent areas:

1. Traffic produced by residents of the neighborhood.[3]
2. Traffic attracted to residences in the neighborhood—visitors, domestic employees, and so forth.
3. Traffic generated by nonresidential land uses which may exist in the neighborhood.
4. Sequential trips to service the area—such as postal and other deliveries, garbage pickup, police patrols, local bus or paratransit services.
5. Traffic generated by occasional activity within the neighborhood—construction projects, utility repair, emergencies and so forth.
6. Traffic generated by land uses in adjacent areas where parking supply does not meet the demand; this traffic spills over into the neighborhood seeking parking opportunities.

Local Traffic Quantities. The first two categories of traffic enumerated above represent trips generated by residential land use activities within the neighborhood. Table 2.2 presents trip generation rates and regression equations from a commonly used U.S. source (Institute of Transportation Engineers, 1987, Section VIII-200), which can be used to obtain a rough estimate of the number of trips likely to be produced and attracted; these data should be used with caution, since many assumptions are concealed behind the figures being quoted. (Other literature on the state of the art of trip generation is contained in many standard texts.)
If time and funds permit, a survey of a sample of households—either within the area being studied or in one representative of one being planned—can be conducted to obtain trip generation and attraction factors.

---

[3] The terminology used here is that of trip generation theory: "production" is trip generation at the home end of home-based trips and the origin of nonhome-based trips; "attraction" is trip generation at the nonhome end of home-based trips and the destination of nonhome-based trips.

TABLE 2.2:

Weekday Vehicle Trip Generation for Residential Areas

Dwelling Type	Avg.	Range	Regression Equation
Trips per Dwelling Unit (DU)			
Single Family Detached Homes	10.1	4.3-21.9	$2.6 + 0.94\, \ell_n$ (DU) (R^2 = .959)
Apartments	6.1	0.5-11.8	$51.0 + 5.92$DU (R^2 = .932)
Residential Condominiums	5.9	0.6-11.8	$2.6 + 0.84\, \ell_n$ (DU) (R^2 = .820)
Occupied Mobile Homes	4.8	2.3- 10.4	No equation available
Trips per Person (P)			
Single Family Detached Homes	2.5	1.2 – 5.2	$-44.0 + 2.55$P (R^2 = .991)
Apartments	2.8	1.2 – 5.8	$252.0 + 2.43$P (R^2 = .952)
Residential Condominiums	2.5	1.1 – 6.7	$310.0 + 1.74$P (R^2 = .925)
Mobile Homes	2.4	1.2 – 3.6	$0.99 + 0.98\, \ell_n$ (P) (R^2 = .858)
Trips per Vehicle Owned (VO)			
Single Family Detached Homes	6.5	1.0 – 9.4	$1.88 + 0.99\, \ell_n$ (VO) (R^2 = .946)
Apartments	5.1	2.9 – 8.6	$-148.0 + 5.48$VO (R^2 = .927)
Residential Condominiums	3.3	1.9 – 7.3	$309.0 + 2.30$VO (R^2 = .915)
Mobile Homes	3.4	1.9 – 4.8	$0.92 + 1.06\, \ell_n$ (VO) (R^2 = .926)

Source: Institute of Transportation Engineers, 1987.

Traffic generated by nonresidential land uses within the neighborhood (the third category of traffic listed above) can be estimated by use of the same data source (Institute of Transportation Engineers, 1987, Section VIII), which includes information on many types of land use or, again, by conducting studies at the site in question or at one similar to the one being planned.

Sequential trips to service the area (category 4), while important when considering street layout, are negligible in quantity, as are those generated by occasional special activities (category 5). In most estimates of trip generation, these traffic sources are ignored; a local traffic count would, of course, include them.

Estimation of overflow traffic (category 6) involves calculation of the parking demand of traffic generators abutting on the neighborhood and estimation of the share of that demand which is likely to spill over into the neighborhood in question. In estimating this spillover, the available spaces at or near the generator must be categorized by those restricted to specific users and those generally available to any employee, customer, or visitor. The analysis also must compare parking fees charged at the traffic generator to free curb parking in the neighborhood—in other words, the tradeoff between parking fees and extra walking distance or other access costs (such as cruising to find a vacant space).

Occasionally, a residential district with only a few points of vehicle access and fairly homogeneous types of households can be used to obtain the sum of trip generation rates for all traffic categories by counting volumes entering and leaving the district. This does not lead to a full understanding of the types of trips being made, but can serve as a check against other methods of estimating trip generation patterns.

Modal Choice

One factor causing the dispersion in vehicle trip generation data shown in Table 2.2 is that of modal shares. Where factors which attract trip makers to transit are present—such as low vehicle ownership and/or disposable income, good transit service, high parking costs at the nonhome trip end—the rate of vehicle trip generation will be somewhat lower than where these factors are absent. Similarly, environments which make it feasible for some trips to be made on foot or by bicycle—areas with relatively high densities and mixed uses for the former, gentle topography and good climate for the latter—generally will exhibit lower vehicle trip rates. It is not possible in this book to deal with the subjects of modal choice characteristics and methods of attracting passengers to transit service; the reader is referred to such literature as Hutchinson (1974) and Lago et al. (1980) for this type of information.

It must be emphasized, however, that quality of transit service to the neighborhood and the presence or absence of such amenities as bike paths, sidewalks, and/or footpaths, are factors which may affect vehicle traffic within that neighborhood. As will be mentioned in subsequent chapters, this should be kept in mind when the layout and design of street networks is considered.

REFERENCES

APPLEYARD, DONALD, and MARK LINTELL. "The Environmental Quality of City Streets: The Residents' Viewpoint." *Journal of the American Institute of Planners*. v. 38, n. 2. March 1972.

APPLEYARD, DONALD. *Livable Streets*. Berkeley, CA: University of California Press, 1981a.

APPLEYARD, DONALD. "Place and Non-Place: The New Search for Roots." In: deNeufville, J. I. (Ed.), *The Land Use Policy Debate in the United States*. New York: Plenum Press, 1981b.

BAKKER, J., and K. HAVINGA. "Helping Pedestrians in Urban Areas." *Papers from the 12th International Study Week in Traffic Engineering and Safety*. London: World Touring and Automobile Association (OTA),1974.

BENNETT, G.T., and R. LANE. "Helping Pedestrians in Urban Areas." *Papers from the 12th International Study Week in Traffic Engineering and Safety*. London: World Touring and Automobile Association (OTA), 1974.

BRINDLE, R.E. "Local Street Traffic and Safety: A Perspective." In: Brindle, R.E. and K.G. Sharp (Eds.), *Local Street Traffic and Safety; Workshop Papers and Discussions*. Vermont South, Australia: Australian Road Research Board, Research Report ARR No. 129, 1983, pp. 1-8.

BRINDLE, R.E., and D.C. ANDREASSEND. "Where Do Reported Bicycle Accidents Occur?" In: *12th Australian Road Research Board Conference, Proceedings,* Part 7, pp. 96-109. Vermont South, Australia: AARB, 1984.

BUCHANAN, COLIN. *Traffic in Towns*. London: Her Majesty's Stationery Office, 1963. (Also published in a "specially shortened edition" by Penguin, 1964.)

FOOT, HUGH C., et al. "Pedestrian Accidents: General Issues and Approaches." In: Chapman, A.J., Wade, F.M. and Foot, H.C. (Eds.) *Pedestrian Accidents*. New York: John Wiley & Sons, 1982.

GANS, HERBERT. *The Levittowners: Ways of Life and Politics in New Suburban Communities*. New York: Pantheon, 1967.

HESTER, RANDOLPH T., JR. *Neighborhood Space: User Needs and Design Responsibility*. Stroudsburg, PA: Dowden, Hutchinson & Ross, 1975.

HUTCHINSON, BRUCE G. *Principles of Urban Transport Systems Planning*. Washington, DC: Scripta Book Co./McGraw-Hill,1974.

INSTITUTE OF TRANSPORTATION ENGINEERS. *Recommended Guidelines for Subdivision Streets*. Washington, D.C., 1984.

INSTITUTE OF TRANSPORTATION ENGINEERS. *Trip Generation (Fourth Edition)*. An Informational Report. Washington, D.C.,1987.

KRAAY, J.H., MATTHIJSSEN, M.P.M., and F.C.M. WEGMAN, *Da verkeersveiligheid in woonwijken*. Leidschendam, the Netherlands: Stichting Wetenschappelijk Onderzoek Verkeersveiligheid, Publicatie 1982-1N, 1982.

LAGO, ARMANDO M., et al. *Patronage Impacts of Changes in Transit Fares and Services*. Washington, D.C.: U.S. Urban Mass Transportation Administration, Office of Service and Methods Demonstrations, 1980.

MUMFORD, LEWIS. *City Development*. New York: Harcourt, Brace & Co., 1945.

NATIONAL SAFETY COUNCIL. *Accident Facts*. Chicago, 1985.

ORGANISATION FOR ECONOMIC CO-OPERATION AND DEVELOPMENT (OECD). *Traffic Safety in Residential Areas*. A report prepared by an OECD research group. Paris, 1979.

OECD. *Traffic Safety of Children*. Scientific Expert Group on Traffic Safety. Paris, 1983.

PEARSON, SIDNEY VERE. *London's Overgrowth and the Cause of Swollen Towns*. London: C.W.Daniel Co., 1939.

PERRY, CLARENCE A. "The Neighborhood Unit." In: *Neighborhood and Community Planning*. Regional Plan of New York and Its Environs. New York, 1929.

PFUNDT, K., and H. HÜLSEN. "Verkehrsunfälle in Berlin." In: *Mitteilungen der Beratungsstelle für Schadenverhütung des H.U.K. Verbandes*. Köln, Germany, 1977.

PICK, FRANK. *Britain Must Rebuild; A Policy for Regional Planning*. London: Kegan Paul, Trench, Trubner & Co., Ltd., 1941.

REPS, JOHN W. *The Making of Urban America: A History of City Planning in the United States*. Princeton, NJ: Princeton University Press, 1965.

SAN FRANCISCO, CITY PLANNING DEPARTMENT. *Transportation: Conditions, Problems, Issues*. San Francisco, 1971.

SNYDER, M.B., and R.L. KNOBLAUCH. *Pedestrian Safety: The Identification of Precipitating Factors and Possible Countermeasures*. Washington, D.C.: U.S. Department of Transportation, Operations Research Report FH11-7312, 1971.

SWOV. *Verkeersveiligheid in woongebieden, wat doen we er aan?* Leidschendam, the Netherlands: Stichting Wetenschappelijk Onderzoek Verkeersveiligheid (SWOV), 1980.

UNTERMANN, RICHARD K. *Accommodating the Pedestrian: Adapting Towns and Neighborhoods for Walking and Bicycling*. New York: Van Nostrand Reinhold Co., 1984.

VERKEER EN WATERSTAAT. *Verblijfsgebieden*. Den Haag, the Netherlands: Directie Verkeersveiligheid van het Ministerie van Verkeer en Waterstaat, 1982.

WADE, FRANCES M., et al. "Accidents and the Physical Environment." In: Chapman, A.J., Wade, F.M. and Foot, H.C. (Eds.) *Pedestrian Accidents*. New York: John Wiley & Sons, 1982.

WEBBER, MELVIN M. "Place and the Non-Place Urban Realm." In: Webber, M.M. et al. (Eds.), *Explorations into Urban Structure*. Philadelphia: Pennsylvania University Press, 1964.

3
Planning for Traffic Control

PLANNING OBJECTIVES FOR RESIDENTIAL STREETS

Residential streets provide a major part of the fabric of cities. How well they function can determine the quality of a city, its safety, comfort, and convenience and the well-being of its citizens. Planning for residential streets is thus a critical government responsibility. Such planning is based in government's duty to protect the public health, safety, and welfare, as well as in its interest in preserving public and private investments and in strengthening the social networks that successful residential neighborhoods support.

Residential street design and traffic control, therefore, should serve neighborhood protection and quality of life objectives. Residential streets should:

- Permit comfortable and safe pedestrian and bicycle movements as well as motorized vehicular movements, and protect vulnerable users such as children, the disabled, and the elderly.
- Accommodate convenient and efficient pickups and deliveries, emergency access (fire, police, ambulance), and maintenance services, and—where densities justify—bus or paratransit services.
- Enhance the overall aesthetics of the neighborhood through well-designed street layout and street landscaping.

The primary function of many residential streets is to provide access to abutting trip origins or destinations, not to provide through movement; their planning and design should proceed accordingly. But collectors and major streets also are home to residents in many communities. Along these streets, the relationship of street space to buildings, and such features as setbacks, landscape treatments, building orientation, construction materials, and ground floor uses can be major determinants of whether residents will consider traffic a serious concern. There are many cases of multiple dwelling units along streets with heavy traffic where appropriate building design and construction have been used and residents are not troubled by traffic. Similarly, there are cases where single-family homes are located along busy streets, but are set far back from the road and protected from traffic by fencing and landscaping. On the other hand, there are many examples in which land use and street use are poorly matched, for example, where modest homes line a busy street and are assaulted daily with noise, fumes, and other adverse traffic impacts. Thus resident traffic problems can and do arise on collectors and major streets, and may require special efforts to balance residents' needs with those of the traveling public.

Hereafter in this chapter, when we speak of residential streets we will mean local streets unless we specifically state otherwise. This leads us to set forth some additional principles about the appropriate uses of local residential streets:

- (Local) residential streets should be protected from through traffic; vehicles traveling on these streets should have a trip origin or destination in the area served by them.
- Residential streets should be protected from vehicles moving at excessive speeds (greater than 25 to 30 mph (40 to 45 km/hr)).
- Residential streets should be protected from parking unrelated to residential activities.

In order to attain these results, the following planning concepts apply:

- Street layout, design, and control should express and reinforce street function.
- The overall street network should include streets designed to accommodate through traffic, as well as residential streets.
- Residential streets should be linked to traffic-carrying streets in a way that simultaneously provides good access to other parts of the community and region and minimizes the chances of the residential streets' use by through traffic.
- Land uses along streets intended to carry through traffic should be selected and designed to minimize their sensitivity to adverse traffic impacts; when possible, uses that can benefit from the greater accessibility and public exposure that major streets can provide should be the ones located on such streets.
- Strategies for reducing auto dependence both by residents and by others are legitimate tools of traffic management for residential streets.

Applying these planning concepts may raise questions about the role of transportation professionals and others in the processes through which traffic management plans are initiated, developed, approved, and implemented. We take the view that residents have a right to livable neighborhoods, and that transportation professionals have an obligation to help assure that streets are designed and managed to support that end. More specifically:

- Residents in existing neighborhoods should have a say in the design, function, and operation of the streets on which they live; they should be able to participate in planning for more livable neighborhoods and have a right to make their preferences known to decision-makers. At the same time, other interests often will be affected by residential traffic controls—business and industry, commuters, even neighboring communities. These interests also should be considered in, and given access to, the planning and design process.

- Transportation professionals—city planners, traffic engineers, others—have a responsibility to identify problems, help formulate alternative solutions, and identify their impacts both on specific affected interests and on the community as a whole; they also can rightfully be expected to offer recommendations to elected officials based on their technical studies and experience.
- Elected officials have the right and the responsibility to determine policies, establish priorities, and make choices concerning traffic planning and management for the community. They often delegate limited authority for these actions to staff, commissions, or even ad hoc groups, but it is the elected officials who are the ultimate decision-makers on major issues.

Planning for residential traffic control thus involves detailed design and regulation of streets; coordinated land use and circulation planning, and broad strategies for reducing auto dependence and supporting the use of alternative transportation modes. This planning occurs in two substantially different situations. In one, planning occurs before substantial amounts of development take place. In such a situation technical issues of design and layout predominate; planning is for the welfare of future residents and users of the area and for their successful integration into the existing community. The emphasis can be on creation of plans and programs that will keep traffic problems from developing while at the same time providing for convenient access and mobility. In the other situation, planning for traffic control occurs in already developed areas. The basic community infrastructure—streets and highways, commercial and residential districts—may be largely in place, and problems may have arisen because of unforeseen changes or poor planning in the past. Controversies over these problems, or over proposed responses, may have developed, and interests may be polarized. Obviously, planning for traffic control is more complex and more difficult in the latter situation. It therefore is advisable to plan ahead and prevent neighborhood traffic problems from developing whenever possible; in cases where problems are already apparent, it is important to have clear policies and procedures in place to guide corrective actions.

The material presented in this chapter is intended to be of use in either set of circumstances, though the applicability of specific strategies and procedures will vary.

The following section provides a review of the most common approaches to land use and transportation planning at the local level, since it is within that context that planning and design for traffic control occurs. The presentation covers the local General or Master Plan; controls on land development, including zoning provisions, subdivision controls, and building permissions; and various transportation-based options including street classification systems, parking codes, and traffic regulations. Then, alternative procedures for traffic control planning are discussed, including citizen roles, the role of local government staff, and the role of elected officials. In the final section, issues of process management (including financing, timing of improvements, and follow-up studies) are considered.

THE PLANNING CONTEXT

Planning for traffic control is carried out in most U.S. communities under a variety of codes, ordinances, and regulations stemming from "police powers" designed to protect and enhance the public health, safety, welfare, morals, and quality of life. These police powers, reserved to the states by the Federal Constitution, are generally delegated in turn to local jurisdictions, or in some cases are reserved for local governments under state constitutions. While there are considerable differences in interpretation from state to state, police powers generally are considered elastic, evolving to accommodate changing community values and concerns. However, the use of police powers is constrained by constitutional principles of equal protection and due process, including constraints on the unlawful taking or damaging of property rights. Furthermore, state legislation frequently shapes the manner in which police powers may be exercised. Thus, most jurisdictions must abide by state-level procedural and substantive requirements in fashioning traffic controls, and they may be limited in the options that can be pursued by state-level restrictions of authority.

It is beyond the scope of this book to present the current situation in each state, and it is noted that outside the United States, legal differences may be substantial. Readers are urged to refer to the government codes and court decisions in effect in their jurisdictions. However, for nearly every U.S. community and jurisdictions in many other countries, three sources of planning guidance will apply and will provide considerable direction for the design and management of neighborhood streets and traffic: (1) a local General or Master Plan; (2) land development regulations; and (3) state and local traffic codes and ordinances (including speed laws, truck ordinances, and parking regulations). Each of these sources of guidance will be discussed (in its U.S. variety) in turn.

The General Plan

From its beginnings at the turn of the century, through its widespread adoption in the decades following the 1928 Standard City Planning Enabling Act, through the expansions and requirements for additional detail of the post-World War II period, the General Plan has come to be the centerpiece of formal land use and circulation planning in most cities and in many counties across the United States. As such, the General Plan constitutes a critical starting point for the planning of residential streets and for traffic control thereon.

General Plans, or Master Plans as they also are called, have been defined by T. J. Kent, Jr. in his classic work on the subject (1964) as

> the official statement of a . . . legislative body which sets forth its major policies concerning desirable future physical development; the published general-plan document must include a single, unified general physical design for the community, and must attempt to clarify the relationships between physical development policies and social and economic goals.

Kent goes on to state that the fundamental purposes that the process of developing and maintaining a General Plan is intended to achieve are

> (1) to improve the physical environment of the community as a setting for human activities—to make it more functional, beautiful, decent, healthful, and efficient; (2) to promote the public interest, the interest of the community at large, rather than the interests of individuals or special groups within the community; (3) to facilitate the democratic determination and implementation of community policies on physical development; (4) to effect political and technical coordination in community development; (5) to inject long-range considerations into the determination of short-range actions; and (6) to bring professional and technical knowledge to bear on the making of political decisions concerning the physical development of the community." (pp. 25-26; punctuation changed)

Seen in such a light, the General Plan offers both a policy framework and a procedural context for coordinating land use and transportation planning, and thus for managing traffic in the city as a whole as well as in its residential areas.

Despite these high ideals, General Plans have been widely criticized as impractical, misleading, and downright irrelevant, often with good reason. In many cities, the General Plan exists as a formality that is readily ignored in day-to-day decision making. In some cases the General Plan is changed so often that no one, not even the planners responsible for it, know exactly what it says at any given moment. Elsewhere the General Plan has been revised so infrequently, despite major changes in the economy, population, and tastes of the community, that it has become outmoded and useless. Some General Plans are little more than wish lists, untested against local data and unsupported by implementing regulations. Some have been adopted piecemeal, with inconsistent elements and conflicting proposals left unresolved.

These problems are all too common, but they are not insurmountable. First, with sound technical work and political commitment to back it up, General Plans can be made to attain their full potential as guiding statements of the community's intentions concerning physical development, and as such they can be used as tools for avoiding many traffic problems and correcting others. Second, General Plans can provide the basis for the development of more detailed ordinances, regulations, and action plans for traffic management, by articulating policies on neighborhood preservation and by setting priorities and standards for circulation planning. When a sound General Plan is in place, transportation professionals can follow its guidance; when the General Plan has serious shortcomings, they can develop recommendations for its revision and strengthening.

The manner in which a General Plan can be used in guiding traffic management for a particular community depends in part on state law. In most cases General Plans must be developed in accord with state legislation defining their scope, content, and application as well as the procedures by which they are to be formulated, adopted, and implemented. Considerable variation can be found in the requirements from state to state, and the broad discretion given to local jurisdictions in many states further results in considerable community-to-community differences. The subject areas that must be covered in the plan, and the relationship of the plan to implementation tools such as zoning and subdivision controls, are generally matters of state policy. Two trends can be identified, however. The first is a trend toward greater breadth of coverage; for example, in recent years there have been moves to include economic, social, and environmental elements as well as physical development elements in the General Plans of a number of communities. Second, there has been a trend toward inclusion of implementation mechanisms within the plan. And while many General Plans traditionally were of "more inspirational than legal effect" (California Office of Planning and Research, 1980), legislative requirements and court decisions that zoning and subdivision approvals be consistent with the General Plan have brought local officials under increasing pressure to adopt precise proposals and clear standards for determining consistency, and to link implementation programs (including capital improvement programs and annual budget cycles) to the General Plan.

While considerable variation in General Plans results from differences in legal context and professional practice, most General Plans in effect in the United States include both a land use element and a circulation element. The land use element typically designates the distribution, location, and extent of the uses of land for housing, business, industry, open space, education, public facilities, and other categories of public and private uses, and may also set forth standards for population density and building intensity. The circulation element typically identifies the general location and extent of streets and highways, transit routes, terminals, parking facilities, and other significant transport facilities, and may set forth standards for their design and operation. In some jurisdictions, other elements may be included in the General Plan that have a direct bearing on traffic planning and management; for example, some states mandate a noise element (requiring that attention be given to protecting sensitive receptors such as housing or hospitals from traffic noise) and/or a conservation element (which may require attention to transportation energy use and/or to residential exposure to automotive pollutant emissions).

Even the simplest of General Plans thus can provide several opportunities for control and management of traffic. First, the plan can provide the framework for assuring that development proceeds in accordance with well thought through and integrated land use and circulation elements. A well-conceived General Plan can be the first step in seeing that land use locations and intensities will not overwhelm the transportation system, and in making a commitment to the development or redevelopment of a circulation system that is able to perform well for the desired location and intensity of land uses. Second, the preparation or updating of a General Plan offers transportation professionals the dual opportunity to identify circulation problems and to propose physical improvements that hold promise for correcting those problems. Third, the General Plan can establish policies that later can be used to guide the development of strategies to resolve neighborhood traffic problems.

A starting point for reviewing the General Plan is to gauge the consistency of the land use and circulation elements. In practice, several problems often arise because the land use and circulation elements of the General Plan are not well coordinated. Common problems include: mismatched land use and transportation elements, such that permitted densities and intensities of development, if fully realized, would overwhelm the ability of available and planned transportation systems to provide an acceptable level of service; designation of certain streets as collectors or major streets when either street design is inadequate for the implied traffic levels, or abutting land uses would be adversely affected by such traffic levels, or both; and inadequate planning for certain transportation modes, including transit, bikes, and other alternatives to the auto, and/or for truck circulation and loading. Each of these problems will be discussed in turn.

Mismatched Land Use and Circulation Elements. Land use and circulation elements can be considered mismatched if the development permitted would overwhelm available and proposed transportation facilities, or if the size, configuration and location of transportation facilities do not correspond to what is needed for efficient circulation and access. This common problem may arise in a number of ways, including adoption of land use and transportation elements at separate times with little or no cross-checking; failure to estimate the trip generation implications of the land use element; overly optimistic forecasts of future transit use and ridesharing; and a tendency on the part of many local governments to develop plans permitting levels of development far in excess of what realistically can be expected. In the latter case, of course, implementation of a transportation plan matched to an unrealistic forecast of land development would result in wasted resources. In other cases, however, development may indeed occur at levels that far exceed the ability of the transportation system—streets, parking, transit—to perform at acceptable levels of service. The result can be serious congestion on the facilities designated to accommodate the generated traffic and parking, resulting in spillovers into residential neighborhoods.

Two obvious alternatives in such cases are to (1) modify the land use element to better reflect the carrying capacities of the transportation facilities; or (2) modify the circulation element to provide sufficient transportation capacity without disturbing residential areas. A third option might be to implement physical and regulatory controls that prevent traffic and parking from entering residential districts, forcing commuters to accept a worsened level of service on the streets and in the parking facilities designated for their use. (Note that in some instances such restrictions may encourage commuters to switch to transit or carpooling, and/or may lead to less development or less traffic-intensive uses, because of the worsened access.)

There are difficulties with all of these strategies. First, city officials are often reluctant to constrain development because of forecasted problems in the transportation system, especially since at the level of detail of a General Plan it is often difficult to determine whether in fact the land use and transportation elements are inconsistent; often, that is not apparent until traffic problems start to develop. Second, increasing transportation capacity may not be financially feasible, and in the case of roadway widenings in built-up areas, may have unacceptable social effects as well. But few citizens or elected officials are likely to consider worsening conditions a palatable option, so that a combination of transportation investment and sensible land development policies may be the best bet.

Inappropriately Designated Major Streets or Collectors.   Major street or collector designation is inappropriate if:

- the roadway itself is not designed to carry moderate to heavy loads of through traffic (including trucks and buses), or
- the abutting land uses are harmed by the amount and characteristics of the traffic on the roadway

Inadequate streets may be designated as major streets or collectors because a planned road widening or reconstruction has not taken place, leaving a street with design characteristics better suited to light local traffic to carry traffic loads it cannot handle well. Or an inappropriate designation may occur because little thought was given to matching roadway design characteristics to traffic levels in the first place: often, streets are designated as collectors or major streets simply because of the amount and kind of traffic that they are carrying, or because they provide direct connections between major traffic-generating nodes or between trip generators and major streets and freeways.

When the problem is simply one of roadway or pavement design, the solutions are obvious: repaving or reconstruction should be scheduled. In many cases, spot improvements (such as intersection redesign at bottleneck areas) may be sufficient. In some cases, regulatory actions (including on-street parking restrictions, and heavy truck restrictions) also may provide relief.

As noted earlier, however, incompatible land uses may be present along streets designated as major streets or collectors. This situation has arisen, for example, when the land uses anticipated when the roadway designations were made are not the uses that actually exist—for instance, when a planned replacement of single-family homes with high-rise apartments never materializes, or when zoning allows housing to be developed in an area designated for commercial or industrial use, and the market dictates housing. Or it once again may be the result of simply designating collectors and major streets without regard to the land uses along the streets.

Whatever the reason, a mismatch between street designation and abutting land uses creates a situation that is ready-made for controversies over traffic. Residents of these streets are frequently unpersuaded that they should live with heavy traffic simply because their street has been designated as a collector or major streets. They may oppose design changes to make their streets better able to carry heavy traffic loads and may argue for removal of the collector or major street designation.

Potential solutions to this problem can be devised through either the land use element or the circulation element of the General Plan. If the street design is adequate but the abutting land uses are adversely affected by the traffic levels, one response might be to rezone the properties for uses that would be more compatible with traffic levels. However, this is unlikely to satisfy aggrieved residents. First, as earlier examples have illustrated, market conditions may not support redevelopment to a more compatible use, so that the rezoning is a hollow solution. Second, even if redevelopment is feasible, it may not be acceptable to residents, who usually want the traffic to go away, not to move themselves. And third, a change in use along the affected street might be incompatible with nearby uses or might compete with more highly valued development opportunities elsewhere in the community.

A second solution might be the use of design features to reduce the impact of the traffic on the residences. For example, noise and glare problems might be reduced by installing sound walls and plantings along the roadway, if the right-of-way is sufficiently wide to accommodate such improvements. Retrofits to the buildings themselves are another alternative; for example, soundproofing and double-glazed windows can reduce the reception of traffic noise, while some kinds of shades, blinds, and curtains can block out headlights. Either strategy, however, may have drawbacks from an aesthetic viewpoint. Furthermore, other problems (such as difficulty in backing out of a driveway because of heavy traffic streams) may be extremely hard to solve.

Revising the circulation plan or making changes in the physical design of the problem street may be other options. In some cases, it may be feasible to identify alternative streets better suited from both a circulation and a land use perspective to carry the traffic; rerouting (possibly with some improvements to the newly designated traffic street) and reclassification and protection of the residential street may be possible. In other cases, a careful examination of available traffic-carrying capacities of the circulation system as a whole may reveal that alternate collectors or major streets are already available to handle the traffic; in such instances the residential street can simply be returned to local status. (Actions to prevent its continued use by through traffic may be needed, of course.) Finally, in some instances it may be possible to leave the General Plan unchanged, and proceed with traffic engineering and operations improvements along the affected street that simultaneously allow it to continue functioning as a major or collector street while responding to the problems leading to residents' concerns. For example, reservation of on-street parking for residents or improving intersections to eliminate rush hour backups may be sufficient to satisfy residents of the affected streets.

Inadequate Planning for Certain Transportation Needs. A circulation plan that fails to provide the kinds of streets needed for efficient bus operation in residential areas, that is silent about sidewalks for pedestrians or routes for bicyclists, or that ignores the need for loading areas in commercial districts, can exacerbate neighborhood traffic problems by reducing the feasibility of alternative modes of transportation and adding to congestion on major streets. This problem often arises because the General Plan has focused on broad street classifications without giving explicit consideration to the specialized needs of buses, bikes, pedestrians, or trucks.

While it is inappropriate for a General Plan to include specific details such as loading zones, transit schedules, and the like, it is good practice to designate the streets where bus service is expected, to identify major truck routes, and to plan a bike route system. Major pedestrian paths and routes also are appropriately included in the General Plan. These special-purpose circulation plans should be developed taking into consideration both the appropriateness of the street system and the needs of the abutting land uses, and should be designed to match the travel needs of the persons to be served. (Too often, bike routes have been designated with little regard for connecting major trip generators, for instance.)

It is equally important to consider the ability of particular land use densities and locations to be served by alternative transportation modes. For example, land use planners need to be aware that the very low densities prevalent in many residential districts can be served by transit only at very high subsidy, so that often it is impossible to provide competitive levels of service. Similarly, the complete segregation of residences from services usually necessitates a level of auto dependence that might be avoided if mixed uses were permitted. Since in many residential neighborhoods a large share of the auto traffic is internally generated, traffic management may well include planning of land uses to reduce such auto dependence when feasible.

Avoiding or correcting these three common flaws in the General Plan—mismatched land use and circulation elements, inappropriately designated major streets or collectors, and inadequate planning for certain transportation needs—should help reduce the incidence of neighborhood traffic problems. But, as the preceding discussion implies, it may not always be possible to devise acceptable strategies through the General Plan itself. Implementation often requires that attention be given to specific zoning regulations, subdivision controls, permit conditions, capital budgeting, and special-purpose ordinances that should be consistent with the General Plan, but in most cases should not be a part of it.

Nevertheless, the General Plan can establish the basic policies and priorities for these implementation tools. Planners should take steps to assure that the General Plan:

- Explicitly states the community's official policies concerning traffic and parking in neighborhoods;
- Establishes priorities between potentially conflicting objectives, such as maintaining and enhancing convenient access for the traveling public and minimization of through traffic on residential streets;
- Sets standards for the performance of various types and categories of transportation facilities and services; and
- Sets forth desired relationships between street use and land use.

With such guidance clearly established, planning professionals will be better equipped to develop detailed implementation strategies; and concerned citizens can more easily be assured that the response they are getting from staff is equitable and even-handed.

To summarize, the General Plan offers several opportunities for avoiding traffic problems, and can also provide the basic framework for developing detailed strategies for managing the problems that do arise. Transportation professionals are advised to review the General Plan for their community and to utilize it as a tool in traffic management planning.

Land Development Regulations

Zoning, subdivision controls, and building permissions are three basic devices for regulating land development. They are in widespread use across the United States and, like the General Plan, are most often employed by local governments in accord with authority granted and requirements established by state legislation. While all three types of development controls preceded General Plans in many areas of the United States, legislation in several states now requires that all be in accord with an adopted General Plan. And while both zoning and subdivision controls often suffer from the same weaknesses and neglect that diminish the utility of many General Plans, they similarly have potential for helping to avoid some traffic problems and for managing others.

Zoning Zoning regulates the type of uses that are permitted and the form of building that can take place. Typical zoning ordinances also control height, size, shape, and placement of buildings on a lot ("bulk" or "envelope" zoning), although modern variations may permit clustering of buildings while maintaining the same overall density across the parcel as a whole. Often, zoning ordinances specify amenities and services which must be provided, such as off-street parking. Zoning thus should be of concern to professionals responsible for traffic management because:

- The permitted locations, densities and intensities of various uses and their relationships to one another are primary determinants of the numbers of trips that will be generated, the modes of transportation that will be used, and the traffic patterns.
- The parking requirements and related services required under zoning provisions may be important in managing traffic and parking in residential neighborhoods.

In this section, both land use and transportation aspects of zoning will be considered.

Land Use Aspects of Zoning: The need to coordinate densities and intensities of various uses with the traffic-carrying capacities of the available transport systems was discussed in the section on General Plans. That discussion will not be repeated here except to note that zoning, as an implementation mechanism for the General Plan, needs to be carefully reviewed to assure that it is consistent with workable streets and livable neighborhoods, and will effectively reinforce the General Plan's policies. Two prevalent land-patterning zoning concepts deserve further attention, however: the segregation of uses through zoning, and the mixed-use concept.

Traditionally, zoning has segregated housing from commercial and industrial development. The reasons for such separation generally have stemmed in large part from a desire to minimize the impact on the resident population of such undesirable byproducts of business and industry as air pollution, noise, and traffic. In addition, the organization of uses into identifiable districts has been viewed as making the provision of needed infrastructure more efficient, producing aggregations that enhance the operation of markets, and increasing overall property values.

Segregation of land uses can be utilized in traffic planning to avoid situations where residential uses will be subjected to heavy flows of commercially oriented traffic. However, problems may arise concerning the zoning of parcels along routes connecting the commercial or industrial zones to one another and to major regional transportation facilities (such as freeways). If the parcels along these routes are zoned commercial, strip development may result. Commercial strips may or may not be considered acceptable from a land use perspective; though they are prevalent in most American communities, they are often considered unaesthetic and are sometimes viewed as inefficient because of the relatively high costs of providing needed public services. In addition, it generally is difficult to develop a pedestrian orientation along commercial strips or to serve them well by transit, with the result that strips are usually best suited to auto-dependent uses and tend to reinforce such auto dependence. The resulting traffic and parking must be carefully managed to avoid spillover into nearby residential areas; physical and regulatory constraints such as traffic barriers, cul-de-sacs, and parking restrictions are sometimes used to accomplish this, as is transition zoning, in which less auto-sensitive uses (such as apartment buildings or open space) are used to buffer areas of single-family dwellings.

If residential uses are permitted along traffic-carrying routes, care must be taken to avoid building in a traffic management problem. Many communities zone land along major and collector streets for apartments and similar higher-density residential uses, rather than for single-family homes. Such a designation may have advantages; higher-density residential developments may need greater street capacity than "local" streets can provide, and problems stemming from traffic can be minimized through careful use of building orientation, floor plans, materials, and other design aspects. For example, the street level or even the first several floors of the building, where traffic impacts tend to be felt most severely, may be devoted to parking or to other relatively noise-insensitive uses such as laundry space, building management offices, and so forth; bedrooms and other noise-sensitive areas can be located away from the street; soundproofing materials in individual quarters can be used to provide protection from outside noises as well as privacy from neighbors. In addition, buildings may be set back from the street, laid out at right angles to the roadway, and buffered with fencing and landscaping to further reduce traffic impact. Depending on state law and local practice, such design features can be mandated directly in the zoning provisions or dealt with in implementing regulations and guidelines, building codes, use permit requirements and conditions, and so forth.

A second problem with segregating residences from jobs and services is that it generally increases auto dependency. While mixed uses are common in older cities, largely the result of uncontrolled or permissive development, recently planners have experimented with the purposeful mixing of complementary uses partly in hopes of reducing this problem. In some cases the mixed use concept simply entails allowing services such as convenience groceries and banks to locate on residentially zoned parcels, usually along collector or major streets but sometimes on the street level of a multiple story building. A more elaborate approach is the Planned Unit Development (PUD) concept, which involves both mixing of uses and (usually) the relaxation of height, bulk, and setback requirements. (PUDs have been pursued in large part out of a desire to keep housing affordable by permitting development of more profitable commercial uses simultaneously.) In either version, mixed-use zoning can shorten some auto trips and replace others with trips by foot or bike; a carefully conceived PUD also can offer some potential for live-work arrangements. On the other hand, there is some evidence from both U.S. and European experience that commercial uses in residential areas may attract trips from all over the community, and that proximate jobs and residences are no assurance that residents will choose to work nearby, and vice versa. Furthermore, mixing of uses requires detailed traffic planning to avoid a situation in which trucks servicing commercial uses become a problem for residential uses of shared street space.

To summarize, the land use-related aspects of zoning should be checked to see that (1) permitted uses, densities, and intensities are consistent with the carrying capacities of the transport systems; (2) permitted uses along connecting routes are appropriately selected and managed; and (3) the traffic implications of mixed uses are throughly accounted for.

Transportation Aspects of Zoning: Local zoning ordinances frequently include several provisions that deal directly with transportation: parking requirements, loading zones, corner controls, and curb cuts are among the most common. (Depending on state law and local practice, these provisions may be found in separate codes or ordinances instead of in the zoning ordinance; or the zoning ordinance may simply establish the legal framework for separate, but binding, resolutions or regulations on these matters.) Transportation-related zoning provisions can be of significance in managing traffic in residential areas in several ways:

- On major thoroughfares, traffic congestion caused by use of travel lanes for loading and unloading can be reduced by the provision of adequate, well-placed loading zones together with enforcement of double-parking prohibition. This in turn reduces the impetus for drivers to divert to neighborhood streets to bypass congestion. Control of parking at corners and control of curb cuts likewise can help reduce traffic tie-ups and otherwise preserve capacity on major streets.
- In commercial districts, an adequate supply of off-street parking both reduces the likelihood that drivers will look for parking in nearby residential neighborhoods and may in some instances preserve the availability of street space for carrying traffic rather than for storing cars.
- In residential areas, adequate off-street parking is a convenience for residents. (However, it has been noted that many residents use street parking in preference to their own garages or other off-street spaces.)
- Finally, control of nonresident parking in residential areas (generally accomplished through resident permit parking programs) can both protect the residential neighborhoods and help to assure the effectiveness of parking policies and transportation system management measures in commercial districts.

The two provisions that will be treated in detail in this chapter are off-street parking requirements and resident permit parking, since they are not only the most complex items in the above list but also offer the greatest potential for contributing to overall traffic management.

Off-Street Parking Requirements: Requirements for off-street parking are in widespread use throughout the United States. Ideally, the number of parking spaces required should be sufficient to meet demand but not so excessive that large numbers of spaces go unoccupied, a circumstance that would both waste valuable urban land that might be better put to more productive use and unnecessarily raise the cost of land development. Thus, parking requirements should reflect the physical and functional characteristics of the buildings the parking is intended to serve, as well as the physical and social characteristics of the urban districts in which they are located. Both use and size of the buildings should be considered (trip generation rates), along with such factors as time of day of use and expected turnover rates (duration of occupancy). In addition, the parking requirements should reflect the existing and anticipated rates of use of transit, ridesharing, bicycling, and walking modes of access.

In practice, many parking provisions of local zoning ordinances fall short of these ideals. As early as 1952, a study sponsored by the Eno Foundation (Mogren and Smith, 1952) noted that many zoning ordinances had been drafted "with minimum investigation of parking requirement characteristics of different building types and land uses," with the result that the requirements were "wholly unrelated to the physical and social characteristics of the cities they are intended to serve" (p. 30). Despite considerable work done by the Eno Foundation, the Institute of Transportation Engineers, the Urban Land Institute, and others in the ensuing years, many cities continue to operate under parking requirements that have not been evaluated against local conditions, and thus require either too little or too much parking for the operating context. Thus, review and refinement of parking requirements may be a valuable step in planning for traffic management; adjusting the requirements to match local conditions and opportunities may both help reduce the chances of spillover into residential neighborhoods of parking and the traffic associated with it, and avoid the possibility that a vast supply of parking will tend to undermine efforts to reduce auto use.

Many communities have found that their parking requirements had been calculated assuming that everyone would drive alone to each site. In revising their off-street parking requirements, a growing number of communities have attempted to account for the availability of alternative modes (especially transit and ridesharing). Some communities have simply lowered the parking requirements to reflect actual mode shares. Others have set goals for transit use and vehicle occupancy that are somewhat higher than existing rates and have adjusted the parking requirement accordingly in hopes of encouraging additional mode shifts. However, this strategy is not effective when an alternative supply of parking on or off the street is available.

Permitting shared parking when there are multiple uses that exhibit substantially different trip peaking and duration is another way that an excessive supply of parking can be avoided. The parking requirement most often is calculated to reflect the peak joint demand rather than the sum of the peak demands without regard to time of use (ULI, 1982).

In a few communities, shared parking takes on a somewhat different meaning: these communities look at the total supply of parking available within a particular district and adjust requirements for additions to the parking supply to make optimal use of spaces over the entire area, rather than building by building. For example, if public facilities have available spaces, those spaces are taken into account in calculating how much additional parking new developments should provide. Often, such provisions are accompanied by requirements for contributions to the costs of operating the off-site spaces in lieu of construction of new spaces.

Another in-lieu strategy permits developers to make commitments to subsidize transit users or suppliers, provide bicycle facilities, or support other transportation systems management (TSM) alternatives as a substitute for a portion of the parking that otherwise would be required. Ordinances permitting such tradeoffs are in effect in Sacramento, Los Angeles, and Seattle, for example. Most of these ordinances are quite new, and there is limited data on their effectiveness. However, experience to date suggests that developers may not take advantage of an optional parking-TSM tradeoff if the procedure for obtaining it is time-consuming and cumbersome (Los Angeles). There also is evidence that under some circumstances, limiting the supply of parking may simply divert motorists to other parking facilities rather than to other modes. For example, in Berkeley, CA, where the University of California has maintained a highly restricted parking supply for many years, available evidence suggests that some mode shifts have occurred, but a larger effect has been massive spillover of parking into nearby residential neighborhoods which, in turn, is leading to the imposition of resident permit parking throughout the affected areas.

The potential for ridesharing can be reflected in a pared-down parking requirement, especially in areas with aggressive carpool and vanpool promotion programs. Furthermore, cities can encourage ridesharing by requiring that conveniently located spaces be assigned to pool vehicles, and/or by mandating free or discount parking for poolers in those cases where there is a parking charge. Another strategy that can be used to manage auto use at a site or in a district involves setting the monthly or daily parking rate so that transit is more price-competitive.

All of these strategies can be incorporated into parking provisions of zoning ordinances, or in some jurisdictions they can be implemented through special ordinances or regulations.

Resident Permit Parking: Permit parking is increasingly employed to restrain nonresidents' use of otherwise unregulated street space in residential areas. Some variations actually prohibit use of the spaces by all nonresidents, while others permit time-limited parking by "outsiders." Resident permit parking is particularly popular in areas where off-street parking for residents is insufficient to meet their needs, but it also is used simply to keep nonresident traffic and parking from intruding into residential neighborhoods. As the Berkeley example above suggests, resident permit parking may be necessary to enforce parking supply restrictions or pricing policies; without constraints on the use of free parking in residential neighborhoods, commuters and shoppers often will choose to balance walking distance against out-of-pocket costs or the inconvenience of using alternative modes.

Resident permit parking or other controls of parking on residential streets can be implemented either through the zoning ordinance or by special legislation. Some jurisdictions have found it useful to identify in the General Plan those areas where parking restrictions might be necessary, to authorize permit parking in the zoning ordinance, and to implement it on a neighborhood-by-neighborhood basis through individual ordinances and regulations.

Subdivision Controls. Subdivision controls pertain to the division of undeveloped land (and occasionally, to land undergoing a major change of use, as is happening in some built-up communities as districts formerly utilized for warehousing, heavy industry, and so forth, are recycled and as skipped-over land is brought into development) into lots or parcels. Early legislation was primarily concerned with avoiding confusion in the registry of deeds, and mostly dealt with methods of surveying and demarcating parcels, recording requirements, and procedures for transferring ownership. But in the period following World War II, concerns about pacing growth at a rate the community could absorb and seeking the developer's help in financing needed public improvements came to the fore, and consequently policies proliferated which required dedications of land for public use, mandated the provision or financing of both on- and off-site infrastructure and services, and timed approvals in accord with community growth policies.

Most states in the United States have a well-developed body of law governing the subdivision of land into developable parcels, and the requirements or exactions that may be imposed on developers. Local land development regulations in accordance with these state laws offer major opportunities for implementing transportation provisions that can minimize traffic problems potentially disruptive of neighborhoods. The land development process can be used to manage traffic through a combination of street design, demand management, and neighborhood protection strategies. When land is being developed, these strategies should be among the first ones considered by concerned transportation professionals, because they can help keep many neighborhood traffic problems from developing in the first place.

Traditionally, subdivision regulations have required the designation of streets and highways necessary for adequate circulation of traffic within the area to be divided, and interconnecting it to the community and region. Local practice has varied substantially as to the particulars, with some jurisdictions requiring little more than the dedication of public easements and others exacting full development of all necessary capital facilities. Recently it has become increasingly common for local governments to insist that subdividers provide certain transit- and ridesharing-related facilities, such as bus pads and stops and land for park-and-ride lots, as well as fully developed and landscaped streets.

Subdivision regulations can be important in controlling traffic in residential neighborhoods in several ways. First, a well-conceived street system can segregate through traffic from local traffic and assure that both collector streets and local-serving streets are designed and constructed to standards that reinforce their intended use (see Chapter 4 for examples). Second, attention can be given to layouts that are suitable for bus operations (with appropriate lane widths, pavement strengths, turning radii, and so forth) and to the provision of facilities that permit and encourage nonvehicular travel—bikeways and sidewalks. The latter features of modern circulation systems are particularly important considering that in many cases a substantial portion of the traffic on residential streets is locally generated. Third, subdivision regulations can help reduce the need for vehicular trips by clustering uses or providing services such that walking or biking is a reasonable alternative.

Building Permissions. In both expanding and intensifying communities, additional leverage over street design and traffic patterns is often available through building permissions—use permits, construction permits, and occupancy permits, among others. These regulatory devices are widely used to condition new development on the provision or funding of both on-site and off-site transportation facilities and services. Both provisions for accommodating project-generated traffic and for moderating the amount of traffic generated are in use.

In order to accommodate traffic, many jurisdictions require developers to make intersection improvements in the immediate vicinity of their projects, to undertake street widenings along major routes leading to the development site, and to provide adequate parking. In order to moderate the impact of project-generated traffic, the scope of these exactions has been expanded in recent years, with requirements to actually reduce the amount of traffic generated by the development or to shift some of it to off-peak periods. In residential projects, for example, local governments have required the provision of on-site services such as convenience groceries and recreational facilities, plus rideshare matching, bike facilities, and provisions for bus service. In commercial developments, these strategies also have been required, along with parking reservations for high-occupancy vehicles, subsidies for transit fares or services, and the like. Some jurisdictions also have required new developments to fund programs and projects designed to reduce the potential for traffic to spill over into residential areas; resident permit parking, traffic diverters placed to prevent entry from commercial projects into neighboring residential streets, and street closures all have been required of new developments.

Building permissions thus can be useful tools both in reducing the likelihood of neighborhood traffic problems and/or in addressing the problems as they arise.

Traffic Codes and Ordinances

The previous sections discussed how land use planning and development regulation can be used to prevent neighborhood traffic problems or, if problems do arise, to create a framework and mechanisms for ameliorating them. In this section, we turn to another important source of tools for guiding neighborhood traffic planning: traffic codes and ordinances.

Traffic codes (or laws, where codification has not occurred) are state legislation governing the manner in which traffic may be regulated. Both procedures for developing and implementing traffic controls (required studies, public hearing requirements, enforcement procedures, record keeping, and so forth) and detailed substantive matters (rules of the road, design and use of traffic control devices and signs, etc.) are commonly covered by state codes or laws. Traffic codes/laws thus both offer guidance for neighborhood traffic planning and specify permissible management strategies. Frequently, they also impose certain restrictions on the ways in which traffic may be regulated. For example, speed limits may be a valuable tool in traffic management, but many states also have a minimum speed below which local jurisdictions ordinarily may not establish or enforce regulations. Similarly, state legislation may prohibit the installation of traffic control devices not explicitly authorized.

As was the case for land use law, it is beyond the scope of this book to catalog and review the traffic codes and laws of the various states and countries. Readers are advised to familiarize themselves with the legal framework that applies in their own jurisdictions, and are cautioned that numerous differences do exist. Despite these differences, however, in the United States all state traffic laws and codes authorize local governments to enforce applicable state legislation within their boundaries, and in most cases they also authorize local governments to take such other detailed actions as are necessary or appropriate to further traffic management objectives consistent with the general framework established at the state level. Local governments commonly carry out this charge through the enactment of local ordinances.

Traffic ordinances are acts passed by local governments under general police powers or specific authority delegated to them by the state. Among the many topics typically covered by traffic ordinances are designation of one-way streets and of truck routes, establishment of turn prohibitions, and regulation of vehicle parking and standing. As was noted earlier, the basis for such detailed ordinances may be established in the General Plan's circulation element or in zoning ordinance requirements. In this manner, traffic ordinances serve both a planning and an implementation function.

In many jurisdictions, traffic ordinances also are being used to spell out conditions under which various neighborhood traffic management strategies will be considered, and to establish standard procedures for developing and implementing neighborhood traffic plans. For example, in Berkeley, CA, a detailed ordinance, supplemented with Council resolutions and administrative regulations, sets forth the procedures by which a request for a traffic diverter is to be made, states findings that must be made before a diverter can be considered, lists impacts that must be analyzed by staff, establishes notice and hearing mechanisms for public involvement, and provides for a follow-up evaluation of all installations. Portland, OR, has developed an ordinance that establishes standards for street design and traffic levels for different parts of the city, taking into consideration land uses along the street as well as traffic flow needs. Such ordinances can be useful to transportation professionals responsible for neighborhood traffic planning both because they can help create a rational framework for considering citizen requests and because they can aid in identifying areas where neighborhood traffic problems may arise, so that advance planning can occur.

It is important to note that other local ordinances related to transportation rather than traffic per se can be of use in traffic management planning. In particular, ordinances and related regulations concerning the operation of taxis and paratransit, and ordinances establishing ridesharing, bicycle, and pedestrian programs, can greatly influence the extent to which alternatives to driving alone are offered and used in a community. Some cities recently have revised their taxi ordinances, for example, to permit freer entry, shared ride services, and jitney-style fixed route operations. Other communities have local ordinances to establish city-based commute alternatives programs, requirements for employer-based ridesharing and transit incentives, bike rack installation requirements and bike route development programs, transit user subsidies, and/or shuttle services. While the impact of such initiatives is most often modest, they do have the effect of reducing auto dependence, and they can be valuable in assuring concerned interests that travel alternatives do in fact exist.

PROCEDURES FOR TRAFFIC CONTROL PLANNING

As the preceding sections illustrate, land use and transportation planning and regulation offer numerous options for neighborhood traffic control. But as many transportation professionals can attest, the process through which traffic controls are developed, evaluated, and introduced can be as important as—some might say more important than— the specifics of the plans and regulations themselves. Elements of the traffic control planning process include:

- Identification of a need for traffic control planning
- Assessment of the problem and its causes
- Development of alternative courses of action that could eliminate or reduce the problem
- Prediction of both primary and secondary impacts of the alternatives
- Negotiation about and choosing of a course of action (or actions)
- Development of an implementation strategy, including a financing and maintenance plan
- Evaluation of the in-place performance of the selected action or actions, and the making of adjustments as needed

The way in which these steps are handled can raise serious questions about professionalism, legitimacy, and fairness; the process can be as salutary or controversial as the actions taken. For this reason, each of the steps listed above will be discussed at some length.

Needs Identification. Traffic problems on residential streets can take a variety of forms, ranging from site-specific hazards such as poor visibility, improper embankment, or rough pavement, to annoyances created by a small category of road users such as trucks, motorcycles, or hot-rodders, to broader problems caused by excessive volumes, excessive average speeds, or excessive nonresident parking. Transportation professionals have utilized two approaches in identifying these problems. In some cities, they regularly inventory conditions on local streets and use adopted standards (either locally developed or taken from state or professional organizations' guidelines) to identify conditions in need of attention. In many other cities, however, neighborhood traffic problems are identified primarily when residents complain about them. These cities may have established standards and procedures for assessing the complaint and dealing with it; more often than not, however, the complaints are handled on a case-by-case basis.

Case-by-case handling of complaints about neighborhood traffic problems can work well when the problems are site-specific or occasional. Indeed, it often is unreasonable to expect transportation professionals to identify such problems, and it may not always be possible to formulate a policy for responding to them in advance. Consider, for example, a noise and speeding problem caused by a motorcycle racing along a scenic residential roadway at night. Transportation professionals are unlikely to know about such a problem until residents inform them of it. Then, depending on the circumstances, they may be able to alleviate the problem by requesting extra enforcement; or, if the problem cannot be effectively ended by enforcement, they may be able to develop circulation pattern changes or roadway design alternatives that are effective.

In some cases, cities are reluctant to develop full inventories of potential roadway hazards because, once there is official recognition of the potential problem, the city's liability in case of accident may be increased. While this has been given as a reason for not identifying neighborhood traffic problems in advance of the city's ability to rectify them, it clearly amounts to an unfortunate compromise.

When problem identification is left to residents, two difficulties may arise. First, residents may demand a specific action that may not be the best solution to their problem. For example, residents may organize a petition requesting installation of stop signs to alleviate speeding, or seeking traffic diverters to keep nonresident parkers out of the neighborhood. Transportation professionals may then find themselves trying to convince a large group of people that their solution is inappropriate. In these circumstances it is important that the transportation professional not merely oppose the residents' solution, but propose a more suitable response. Otherwise, relations between staff and citizens may deteriorate; residents may become convinced that the staff is unsympathetic and uncooperative, and attempt to have their solution imposed by elected officials.

The second difficulty that may arise when problem identification is left to residents is that complaints do not always come from the areas where city records indicate such problems are worst. Affluent residents, as well as longer-term residents, may be able to organize to put their concerns on the public agenda much more effectively than those who are poorer or more transient. Informal studies in Berkeley and Palo Alto have suggested that homeowners, upper-income residents, and those with higher levels of education are more likely to complain about traffic than those who, according to the data, experience greater adverse impacts. By responding only to complaints, transportation professionals may not always be directing scarce city resources to the areas with greatest needs. On the other hand, few citizen groups will be satisfied by a response that their problem is not as severe as someone else's.

In general, it is advisable for transportation professionals to take some initiative in identifying neighborhood traffic problems throughout the community and in devising approaches for alleviating them. It may be appropriate to establish traffic volume, speed, and composition standards for residential streets, and to identify the General Plan in those areas where problems exist. Portland, OR, has used such an approach and has found it useful both in assessing individual requests and in setting priorities for action. In addition, written guidelines on the use of various traffic control devices can be valuable in advising residents about appropriate solutions to the problems they articulate, and may help both the transportation professional and elected officials in making decisions about resident requests. Cupertino, CA, and a number of other cities have established guidelines on stop signs, traffic diverters, and other control devices. Finally, clear procedures on how resident requests will be handled, including procedures for petitioning, public notice, hearing, technical studies, and decision making can help reduce the potential for misunderstanding and conflict. Berkeley, CA, has written procedural guidelines for initiating studies for diverter installation or removal and for resident permit parking programs.

Assessment of the Problem and Its Causes. Once a neighborhood traffic problem has been identified, it usually is necessary for transportation professionals to carry out field studies to develop a clearer understanding of the problem's magnitude and origins. The specific types of field studies that may be needed will vary with the nature of the complaint, but generally information about traffic volumes, speed, and composition, accidents, street geometrics and other design features, parking supply and use patterns, land use characteristics, and other special conditions (such as concentration of families with children, and heavy bike traffic) will be relevant. Often, a field study will reveal that the source of the neighborhood traffic problem is different from, or at least more complex than, what was originally assumed. For example, on-street parking shortages in certain neighborhoods have been found to be the result not only of commuter use of the spaces, but partly due to residential off-street parking standards being below the level needed to handle residents' cars. Similarly, some communities have found that heavy traffic volumes are the result of heavy auto use by residents of the city, and not simply the product of heavy commuting by employees living elsewhere.

In examining a neighborhood traffic problem, it is important to extend the investigation widely enough to identify the full range of causes. For example, commuter shortcutting on residential streets may be the result of inadequate control of parking and loading on major streets some blocks away. Or nonresident parking on-street in residential neighborhoods may be encouraged by the pricing policies or hours of operation in commercial parking facilities.

Development of Alternative Courses of Action. As the preceding discussion suggests, there often are several ways in which a neighborhood traffic problem might be alleviated. In the case of speeding through neighborhoods, residents may suggest stop sign installation; transportation professionals usually will prefer heavier enforcement, a change in the circulation plan, or installation of roadway undulations. (See Chapter 5 for a discussion of these alternatives.) In the case of locally generated traffic, it may be necessary to create a community awareness program and commute alternatives for residents, in addition to programs directed at employees. In the case of traffic diverting from a congested major street, it may be equally appropriate to implement circulation changes, install devices to prevent or reduce the diversion, or improve traffic flow on the major street (through parking and loading restrictions, better signal timings, and so forth).

In developing alternative courses of action, transportation professionals should consider both short-term options and those that might be implemented only over the longer term. For example, a short-term solution to nonresident parking in neighborhoods might be to implement resident permit parking. Longer-term solutions would include developing commute-alternative programs to reduce the percentage of nonresidents arriving by auto, improving the management of available commercial parking facilities to make space utilization more effective, and constructing additional parking spaces for nonresidents' use. It should be noted, of course, that residents will usually want a quick solution to their problem; they are unlikely to be satisfied by a proposal to alleviate a problem through long-term, gradual change, and may not even be particularly supportive of implementation of the longer-term measures in addition to the short-term ones. Transportation professionals, however, should recognize that a combination of short-term and longer-term solutions may be in the best overall interest of the community.

It is important to include among the alternatives those suggested by community groups, even if the transportation professional believes them to be inadvisable or infeasible. To do otherwise is likely to create unnecessary antagonisms. Deficiencies in proposals are best pointed out in evaluating the alternatives, rather than eliminating the proposals from the start.

Impact Prediction. Once a set of alternatives that could eliminate or reduce the neighborhood traffic control problem has been sketched out, the next step is to assess their probable impacts. The feasibility of the alternatives and the extent to which they are likely to be successful in reducing the problem are of primary concern. When the costs of a proposed course of action make its full implementation questionable, or when implementation would be possible only if new ordinances were enacted or a major new funding source (such as an assessment district) were established, these facts should be explicitly recognized. Similarly, the transportation professional should consider not only the effectiveness that might be possible should full implementation and compliance occur, but also the probable effectiveness taking into consideration the likelihood of partial implementation and/or compliance. Enforcement of speed laws offers a good example of the differences in impact under full implementation and partial implementation conditions. When enforcement is regularly carried out, it may well be preferable to physical controls on speeding. However, in many communities traffic enforcement divisions are short-staffed so that, realistically, enforcement on low-to-moderate-volume streets in residential neighborhoods can be sporadic at best. When this is the case, the transportation professional should recognize it in assessing the effectiveness of relying on enforcement to correct a speeding problem.

In addition to the basic issues of feasibility and effectiveness, other primary impacts that should be assessed include:

- effects on traffic volumes, time of day of travel, and traffic composition
- effects on trip lengths and circuity
- effects on vehicle operations, including stops and starts
- impacts resulting from changes in the above traffic and trip characteristics, including fuel consumption, pollutant emissions, noise impacts, and traffic safety
- impacts on neighborhood quality from the changes in the above characteristics

Many traffic control alternatives will have positive effects with regard to some of these factors, and negative effects on others; for example, installation of diverters may increase trip circuity, stop-and-go driving, fuel use, and emissions, but greatly improve the tranquility of the residential neighborhood and transfer the traffic problems to locations that are less sensitive. Chapter 5 discusses past experiences with various traffic control approaches with regard to many of these impacts.

In addition to the direct impacts on the affected neighborhood and traffic, neighborhood traffic controls may have a variety of wider-ranging effects. Of particular concern are undesirable impacts on other streets or neighborhoods. For example, traffic control devices or parking restrictions in one neighborhood may lead motorists to search out alternative routes or parking in other, unprotected neighborhoods. Or motorists may overload the major streets designated to carry traffic or the parking facilities available for commercial use. Transportation professionals may find it necessary to extend their search for alternatives more broadly to avoid or minimize unintended effects (such as by including several neighborhoods in a resident permit parking program to avoid spillover problems) or to take positive actions to alleviate the adverse effects (as in making traffic engineering improvements on the major street designated to carry heavier traffic loads to maintain acceptable operations).

Other secondary impacts that may arise include effects on transit operations, on emergency vehicle access and response times, and on both routine and occasional pick-up and delivery services. In most cases it is possible to minimize these impacts or to develop acceptable new plans for these services; this should be considered an explicit part of the planning for neighborhood traffic control. Certain other secondary impacts may be more difficult to predict or plan for; for example, certain kinds of neighborhood traffic controls may produce changes in property values in both residential and commercial districts. By improving the quality of residential neighborhoods and taking steps to maintain accessibility to commercial areas, it usually should be possible to assure that these effects will be positive.

Choosing a Course of Action. In the preceding steps the transportation professionals will have developed alternatives for alleviating a neighborhood traffic problem; examined their feasibility, effectiveness, and costs, identified their positive and negative impacts; and developed ways to avoid or minimize major negative effects. In this evaluation process, they generally will state their opinions as to the desirability of the various courses of action. However, deciding what to do is usually the responsibility of elected officials.

Transportation professionals can help these decision-makers in their choice not only by developing a sound, thorough report on the alternatives and their likely consequences, but by assuring that concerned citizens—including those indirectly affected by the various proposals, as well as the smaller group that is directly affected—have been adequately involved throughout the planning process.

A variety of citizen participation techniques are available to help carry out this task. Among those commonly used in transportation planning efforts are:

- maintaining telephone or mail contact with persons who have brought a problem to the staff's attention
- establishing a clear work program and schedule for the necessary planning activities, and announcing the schedule to the public
- maintaining a mailing list of citizen groups, business associations, and so forth, that may have an interest in transportation planning activities, and using the list to inform a broader set of interests about the planning effort
- working with an established commission or committee (planning commission, transportation commission, council subcommittee, and so forth) to review the request and subsequent planning products
- developing interim memoranda or reports on various aspects of the analysis (such as assessment of the problem, alternatives to be considered, and impact assessment), and making these products available for citizen review and comment.
- posting notices of the planning activities and proposed changes in the affected areas (or mailing notices, where such an option is affordable)
- scheduling workshops and public hearings to review the major work products and gather comments and suggestions
- making presentations at community groups' meetings to inform them of the planning activities and solicit input

Clear documentation of these participation efforts can be invaluable to elected officials when the time comes to make a decision. In documenting the participation efforts it is useful to prepare a succinct summary of each meeting, including major questions, concerns and suggestions that arose. A sign-in sheet also can be attached, providing an indication of the extent of public interest in the matter, the groups and neighborhoods represented at the meeting, and so forth. It is of course important that concerns and suggestions presented by the public be addressed in the work carried out by staff.

Developing an Implementation Strategy. Once a course of action has been decided upon, a detailed plan for implementing it can be put together. This generally will include development of any regulations necessary for effective implementation, development of detailed budgets for the various steps to be taken, and field implementation itself. This last step may necessitate public notice, so that both residents and motorists are forewarned that changes are about to take place. Chapter 6 discusses implementation in greater detail.

Evaluating In-place Performance. In many cases, unanticipated effects may result from the implementation of a neighborhood traffic control project or program. It is therefore essential to monitor the implementation and to be ready to make adjustments, if necessary. Generally, it is appropriate to wait for a period of three to six months before making changes, since the earlier period may be one in which both residents and others are adjusting to the new conditions. It is often useful to inform affected interests that this three-to-six-month period of adjustment may be needed, so that they are prepared for minor problems, and to let them know that minor revisions may be made based on monitoring results. Of course, major changes generally should not be made without another thorough planning effort. The evaluation of in-place projects and programs is discussed more thoroughly in Chapter 6.

By following a systematic, participatory process for assessing neighborhood traffic problems and devising and implementing responses, transportation professionals will increase the likelihood that their proposals will be considered fair and equitable. This in turn should increase the chances that they will be accepted by the public and their elected officials, and will be successful in reducing neighborhood traffic problems.

REFERENCES

BAGBY, D. GORDON. "The Effects of Traffic Flow on Residential Property Values." *Journal of the American Planning Association,* January 1980, pp. 88-94.

Bridging the Gap: Using Findings in Local Land Use Decisions. State of California, Office of Planning and Research,Sacramento, CA, December 1982.

DARE, JAMES W., and NOEL F. SCHOENMAN. "Seattle's Neighborhood Traffic Control Program." *ITE Journal,* February 1982. pp. 22-25.

General Plan Guidelines, State of California, Office of Planning and Research, Sacramento, CA, September 1980, (updated periodically).

GREENBERG, S.W., and W.M. ROHE. "Neighborhood Design and Crime: A Test of Two Perspectives." *Journal of the American Planning Association,* Winter 1984, pp. 48-61.

INSTITUTE OF TRANSPORTATION ENGINEERS. *Proceedings: International Symposium on Neighborhood Traffic Restraints,* 1981.

JEPSEN, STEVEN R. "The American Woonerf, Boulder's Experience." *1985 Institute of Transportation Engineers Compendium of Technical Papers,* pp. 102-107.

KANELY, BRIAN D., and BRUCE E. FERRIS. "Traffic Diverters for Residential Traffic Control—The Gainesville Experience." *1985 Institute of Transportation Engineers Compendium of Technical Papers,* pp. 72-76.

KENT, T.J., JR. *The Urban General Plan.* San Francisco: Chandler Publishing Co., 1964.

KOSTKA, JOSEPH V. *Planning Residential Subdivisions.* Winnipeg, Canada: Hignell Printing Limited, June 1954.

KUEMMEL, DAVID A. "A Residential Traffic Management Plan in U.S.A.'s Snowbelt." *1983 Institute of Transportation Engineers Compendium of Technical Papers,* pp. 12-6–12-15.

MARCONI, WILLIAM. "Anatomy of a Failure." *ITE Journal,* March 1981, pp. 26-29.

MARKS, HAROLD. *Traffic Circulation Planning for Communities.* Los Angeles, CA: Gruen Associates, 1974.

MEIER, DIANE. "The Policy Adopted in Arlington County, Virginia, for Solving Real and Perceived Speeding Problems on Residential Streets." *1985 Institute of Transportation Engineers Compendium of Technical Papers,* pp. 97-101.

MOGREN, EDWARD G., and WILBUR S. SMITH. *Zoning and Traffic.* Saugatuck, CT: The Eno Foundation for Highway Traffic Control, 1952.

City of Portland, Oregon. "Neighborhood Traffic Management Process," 1984.

RISER, C.E. "Neighborhood Traffic Development in Toledo." *Technical Notes,* Institute of Transportation Engineers, October 1980, pp. 8-14.

SARNA, JOHN L., and ROBERT HINTERSTEINER. "Community Perception of Neighborhood Traffic Problems." *1980 Institute of Transportation of Engineers Compendium of Technical Papers,* pp. 92-98.

SOUTH AUSTRALIA DIRECTOR-GENERAL of TRANSPORT. *Evaluating Residential Street Management Schemes: Guidelines and Criteria.* Adelaide, South Australia, 1984.

SOUTH AUSTRALIA DIRECTOR-GENERAL of TRANSPORT. *Residential Street Management: First Seminar Report.* Adelaide, South Australia, 1983, p. 102.

URBAN CONSORTIUM FOR TECHNOLOGY INITIATIVES. *The Impact of Traffic on Residential Areas.* Washington, D.C.: U.S. Department of Transportation, June 1982.

WIERSIG, DOUGLAS W., and JOHN W. VAN WINKLE. "Neighborhood Traffic Management in the Dallas/Fort Worth Area." *1985 Institute of Transportation Engineers Compendium of Technical Papers,* pp. 82-86.

4
Design
and Redesign
of Neighborhood Streets

The preceding chapter discussed conflicts between neighborhood activities and automobile traffic. It described planning methods for traffic management and control. This chapter examines the possibilities for street redesign and explains design principles for new residential streets. The goal is to achieve a balance between the nonmotorized street users and the motorized users.

Street design should influence the behavior of the street user in order to promote safety and livability. The designs illustrated in this chapter address two major problems:

- protection of local streets from through traffic, and
- mitigation of traffic impacts on collector or major streets in residential areas

Inappropriately designated major or collector streets are a serious concern in many cities. In San Francisco, a city of only 49 square miles (125 sq km), approximately 80 miles (130 km) of residential streets carry more than 10,000 vehicles per day (Fig. 4.1). On 20 miles (32 km) of such streets, the daily traffic volumes exceed 30,000 vehicles (City of San Francisco, 1981). This problem is widespread, and is not limited to inner city areas. In recently developed subdivisions (Fig. 4.2), some streets have been laid out with four 12-foot wide lanes, two parking lanes, and a median strip. With wide streets in place, arguments for connecting them to other streets and thus forcing them to carry additional traffic can easily be made in the future when more subdivisions are added. Also, wide streets like these can easily be used by motorists as shortcuts or bypasses, avoiding congested major streets or freeway sections.

Some northern European countries have introduced a two-category street classification: streets with traffic preference and streets designed for primarily residential activities.

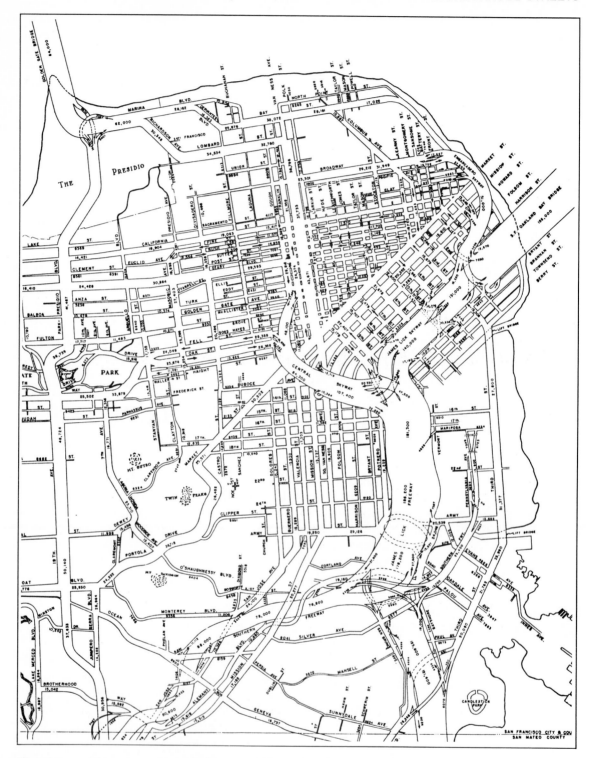

Figure 4.1   Twenty-four hour traffic flow on principal streets and highways in San Francisco.

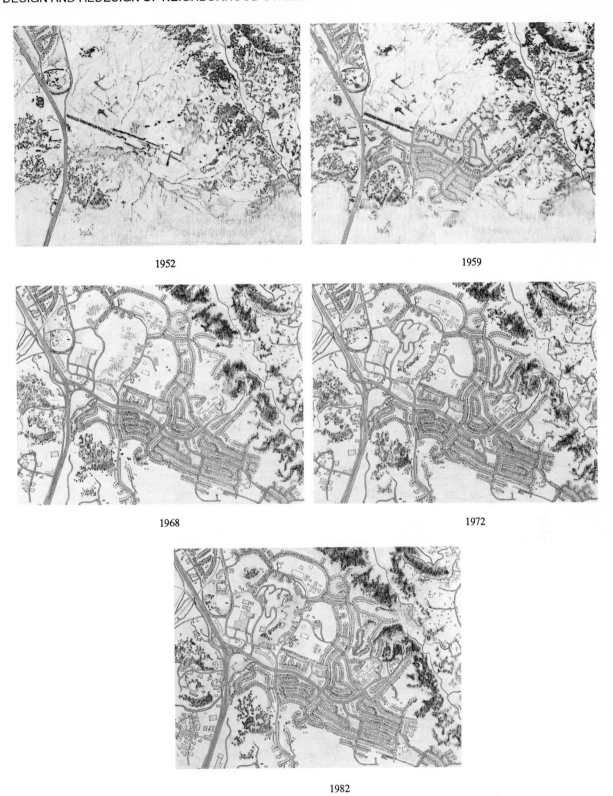

1952   1959

1968   1972

1982

Figure 4.2   The changing suburb. As more land becomes available for development, additional
streets are laid out and linked to east-west collector streets. Ten years after the initial
development started it is still growing, making it necessary to extend existing north-
south streets wherever possible. Sixteen years later, some of these north-south streets
become collector streets.

As long as streets with traffic preference are predominantly residential in character, however, the problem persists.

As early as 1928 planners responded to the incompatibility of residential activities and traffic flows by separating car traffic from pedestrian traffic. The prototype for this separation of circulation was the "New Town" of Radburn, New Jersey, mentioned in Chapter 2. In Radburn, a complete separation of traffic types into fast and slow has been achieved. Homes and other facilities can be reached by car as well as on foot, but on different path systems without ever crossing at grade.

The Radburn system uses the hierarchical road classification system, with cul-de-sacs for all local residential streets. This system has been applied in many countries. In Sweden the so-called Scaft Guidelines establish separation between fast and slow traffic. Similar guidelines were used for "New Towns" in the Netherlands and in Great Britain. However, even this system poses problems. The complete separation of traffic is associated with high costs, and there has been criticism of the poor livability of the residential districts (Gehl, 1986).

The problem of through traffic on local residential streets is easier to solve. A recent invention is the idea of integrating car traffic with all other activities on *one* road surface. Motorized traffic and bicycle traffic is forced to adapt to pedestrian behavior. The "Woonerf," as this type of street is called, was first built in the Netherlands (Fig. 4.3). It has become very popular in other European countries (Fig. 4.4).

Figure 4.3 A Woonerf in an inner city neighborhood.

Figure 4.4 An example of a Danish "slow street."

Extensive installations and utility relocations are required for this purpose, and initial costs may therefore be high. An exception to this is in areas where residential streets, curbs, and sidewalks are constructed of paving blocks bedded in sand. Due to settlement, such blocks are usually taken up and reset at intervals of six years of so. At such a time there is little extra cost involved in resetting the blocks in patterns characteristic of a Woonerf rather than of a normal street—only costs for design, street furniture, plantings, and possible utility relocation are extra.

The higher maintenance cost incurred by such facilities should also be taken into account. These costs are offset by the positive effects on the residential users and on the experience of using such a street as a driver or pedestrian.

Figure 4.5 The residential precinct (Woonerf). Source: Royal Dutch Touring Club. *Woonerf.* The Hague, 1980

Figure 4.6   Woonerf traffic control sign at the entrance to a retrofitted residential street.

The early Woonerven were laid out in the late 1960s. Uniform pavement of the entire right-of-way is a common design feature. Both sidewalks, as well as the street, are surfaced without grade changes. Cars in motion, parked cars, pedestrians, bicyclists, and children at play share the street space. Through traffic is permitted, but street furniture and planting design make it impossible to drive fast. The car driver has to negotiate or pass narrow sections of roadway. The presence of children's play areas, recreation facilities, trees, and flower beds clearly signal to the car driver an entrance into a "residential precinct" scaled to the human dimension (Fig. 4.5). In some Woonerven, the roadway width is no more than 6 feet (2 m) for two-way traffic with widening for passing every 100 feet (30 m) and shifting from side to side every 125 feet (40 m). Changes of route and the suggestion of narrowness of the traveled way are accentuated by pavement pattern contrasts.

One-way operation on Woonerven is not advocated because cars are tempted to drive at higher speeds. At crosswalks where children play, additional narrowing, bumps, and thresholds are used. Parking spaces are designed and limited so that only vehicles of up to 22 feet (6.5 m) by 6 feet (1.8 m) can enter these areas. Actual width between vertical obstacles is always left sufficient for passage at low speeds of garbage trucks, firefighting equipment, and moving vans. Right-angle parking spaces are preferred because they demand more attention from the driver and can be used better by children when they are empty. Parking spaces are limited to clusters of six or seven.

The Dutch Ministry of Transport formally adopted the Woonerf concept in 1976 by including it in the Traffic Code of the Netherlands. A new traffic sign was developed (Fig. 4.6). Placed at the entrance of each Woonerf, it mandates a 15 mph speed limit and gives the right-of-way to the pedestrian. The same sign with a diagonal red bar is used at the exit of each Woonerf.

More recent Woonerven have been designed primarily in newer residential districts. Older districts are sometimes less suitable for establishing Woonerven, because their high population densities and mixed land uses create greater needs for parking space, and because they include many features which attract traffic. Also, their locations often result in a great deal of through traffic which cannot be easily accommodated elsewhere. Residential streets in these districts are usually redesigned as Woonerven only if a neighborhood is part of a redevelopment or preservation district. This also relates to the relatively high costs and the rather stringent requirements a Woonerf must satisfy.

In the Netherlands, about 2,700 residential streets were converted to Woonerven between 1976 and 1983. In a 1981 survey (Vissers, 1982) using a nationally representative random sample, 70 percent of the residents considered the Woonerf desirable or very desirable, 16 percent were indifferent, and 15 percent were opposed. Opposition was chiefly based upon the association of Woonerven with low-income neighborhoods. Indeed, early Woonerven were established in inner city rehabilitation areas.

The generally positive view of Woonerven is not only due to improved safety considerations (OECD, 1979). Woonerven are also perceived as highly desirable because they provide play space for children and parklike areas with the quietness and atmosphere appropriate for a residential precinct. However, problems still exist with speeding moped users, a point which has been confirmed by speed checks and measurements.

The Dutch Woonerf concept has remained a novelty in North American residential communities. In some instances, streets in newly laid out "planned unit" developments have been designed to accommodate foot, bicycle, and motor vehicle traffic on one shared street surface (Figs. 4.7, 4.8). Frequently these streets have been designated exclusively for use by residents of the development. Widespread application of the Woonerf concept will not be likely in North American neighborhoods due to the cost of these improvements. The concept, however, demonstrates that it is possible to influence driver behavior through street design.

Figure 4.7 Shared street surface for cars, pedestrians, and bicycles. (Photo by Joshua Freiwald).

Figure 4.8 The islands: A planned unit development in the San Francisco Bay area. (Photo by Joshua Freiwald).

AREA-WIDE TRAFFIC RESTRAINT PROGRAMS

Experience in West Germany and other countries points to a combination of measures carefully designed for specific situations. In recent years, German cities have taken the lead in the creation of safer residential environments. The idea is based on Buchanan's concept of "environmental areas." The Germans call it "Verkehrsberuhigung," tranquilization of car traffic. It consists of a package of traffic restrictions on motorized vehicles, with the exception of public transport, in favor of pedestrians and bicycle riders.

The package includes an adapted Woonerf design (Fig. 4.9) for local streets and an area-wide 30 km/h (20 mph) speed limit. Obstacles are installed against through traffic. This usually results in longer routes for traffic terminating in residential areas. Through traffic, especially trucks, are discouraged from entering the area, and are diverted whenever possible to nonresidential streets. Intervention in traffic circulation has lowered the density of cars in residential districts.

Changes are also made to collector and major streets. The number of lanes is reduced to a minimum to facilitate the planting of trees and the widening of sidewalks, often with new bicycle lanes. Whenever possible, a change away from straight road alignment is made. Often, street entrances are necked down (Fig. 4.10), and raised or brick crosswalks are built in combination with pavement undulations.

In the city of Berlin, boulevards are currently being rehabilitated with double rows of trees (Fig. 4.11). These wide thoroughfares historically had tree-lined carriageways in the middle of the road. They were later used as streetcar rights-of-way. After the electric trolleys disappeared, the trees were cut down and room was made for additional moving lanes. Today, trees are being replanted. The space under the trees is used for pedestrians, bicyclists, or parking.

Since 1982, two hundred municipalities in West Germany have introduced restraint schemes. Substantial research programs on area-wide environmental traffic management strategies are in progress at the federal level. Pilot programs have been conducted in cities and neighborhoods varying in size from 2,600 to 100,000 inhabitants. Preliminary results of the research show a 60 percent decline in injury accidents. However, the total number of accidents stayed about the same. The important change is a reduction in the severity of accidents (City of Berlin, Department of City Planning and Environment, 1983; Keller, 1985).

Figure 4.9   An intersection designed as a play area.

Figure 4.10   Chokers in the Pimlico district, London.

Figure 4.11   Rehabilitation of an inner-city boulevard with diagonal parking in the center strip (Berlin).

Street redesign has reduced traffic speed. Experience with signage has shown little effect. Only a 5 percent reduction was observed (Keller, 1984). More success was obtained in speed reduction on streets redesigned for lower speeds. Here, a 10 percent drop was observed (Table 4.1).

Traffic volume and noise reduction were measured in Berlin. In a sample of 13 streets, a before-and-after comparison was made. Streets in the traffic restraint program showed an average of 3 to 5 dB reduction in noise, and in some cases up to 9 dB. Table 4.1 shows a reduction in traffic volume of 50 percent on half of the streets in the sample.

TABLE 4.1

SOME EFFECTS OF VERKEHRSBERUHIGUNG IN BERLIN

Street	Motorized Vehicles per hour		Motorized Vehicles percent trucks		Average Noise Level dB(A)		Average Speed (km/h)	
	before	after	before	after	before	after	before	after
Knobelsdorferstr. A	430	220	9	8	70.3	62.7	38	23
Knobelsdorferstr. B	357	213	7	8	68.9	63.9	38	26
Danckelmannstr.	110	120	17	10	65.5	63.2	30	23
Sigmaringer Str.	186	90	11	8	65.6	61.1	37	31
Neue Christstr.	147	138	7	5	65.4	63.2	35	27
Paulsenstr.	217	100	4	4	63.3	60.0	—	23
Wrangelstr.	103	90	13	8	66.9	62.4	35	22
Hildegardstr.	760	330	7	8	70.0	66.0	45	37
Fahrenheitstr.	73	28	6	0	60.1	56.3	—	—
Taborstr. A	175	137	14	11	67.2	63.0	36	23
Taborstr. B	130	117	16	9	67.1	62.8	39	26
Saatwinkler Damm A	517		15		69.6	—	57	
Saatwinkler Damm B		1135		17	—	69.9		34

Source: Giesler, Nolle, 1984.

Influencing Driver Behavior through Design

The European experience is useful for North American transportation planning and design. Indeed, traffic management schemes have been applied in a great number of communities in the United States and Canada, and evaluations of these schemes are being carried out (see Chapters 5 and 6).

However, comparatively little use has been made of street design changes—change to the street space that alters the appearance of the street and changes driver behavior. Such physical changes include landscaping or careful design of curbs, street surfaces, lighting, bicycle paths, crosswalks, and many other features. These design changes can influence driver behavior. They influence the way car drivers and all other users experience the street (Appleyard, Bosselmann, 1980).

Improvements for street design can solve safety problems on residential streets with low traffic volumes. Here the occasional fast car can be very dangerous to playing children, and is perceived as dangerous by all other non-motorized street users, including bicyclists.

Streets as Places. On most streets in North American cities car drivers do not have the experience of entering into places for residential activities. The streets provide long views, allowing the driver to anticipate future events with reasonable certainty. The edges of the street space consist of uninterrupted horizontal lines converging on the axis of the road (Fig. 4.12). Many cars face in the same direction and offer little threat of quickly moving across the driver's path. Frequently there are few signs of residents or children, and, often, few trees are tall enough to interrupt the view. In sum, the streets look like and are experienced as channels designed for the driver.

However, it is possible to create memorable images of places in the minds of drivers. For example, the street shown in Fig. 4.13 has a natural depression framed by tall old trees.

Figure 4.12 Typical local street in a 1950 suburban neighborhood. The street is 36 ft wide.

Figure 4.13 Local street with natural depression. Drivers will slow down because the movement of others cannot be anticipated.

Car drivers will slow down as they approach, because the movement of others beyond the visible street space cannot be anticipated. Similarly, on existing streets, long views can be improved by restraining them (Fig. 4.14). For example, a visual narrowing of the street space at the beginning of a block or in the middle can create a "gateway into a residential place." The long converging horizontal lines are broken by vertical elements such as trees. Through street design, trees can frame new vistas, and at the same time create entrances into streets or visually divide a long street into two shorter ones. On local streets, curbs, which are generally parallel to property lines, can take on a slightly meandering form. With landscaping, a street will not only have a new visual quality but also draw the driver's attention to maneuvering the vehicle through a more complicated street space. This alerts

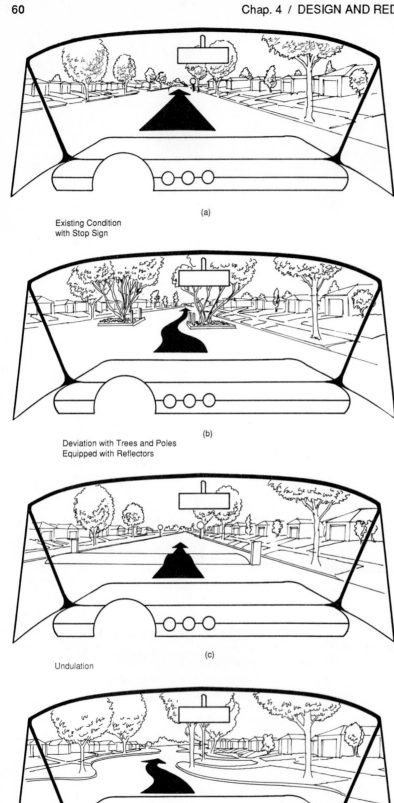

(a)

Existing Condition
with Stop Sign

(b)

Deviation with Trees and Poles
Equipped with Reflectors

(c)

Undulation

(d)

Deviation with Areas for
Additional Landscaping

Figure 4.14 Alternative designs for
a wide suburban street.

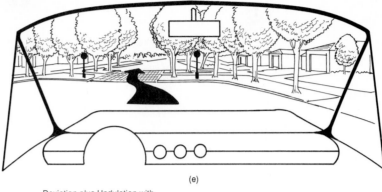

(e)

Deviation plus Undulation with
Raised Crosswalk and Light Fixtures

the driver to look out for other street users. It will slow the speed to a more appropriate travel pace. Designing for the appropriate travel speed takes the psychology of drivers into consideration.

Appearance of the street is a primary concern of street residents, because it is a reflection of their social identity and affects their property values. Street design can enhance the appearance of the street or detract from it. Some temporary traffic control devices used in U.S. cities—painted garbage cans or wooden barriers—have been tolerated for short periods, but can become a source of criticism if they become permanent. An environmental designer or landscape architect should be asked to participate in the design of the devices, even temporary ones, so that they fit in with the neighborhood character.

POLICIES FOR STREET DESIGN

1. *Traffic management devices and changes to the street design should be compatible with the character of the neighborhood.* Attractive paving surfaces, natural materials, vegetation, and trees in keeping with local character are better than guardrails. The warm, light colors and textures of street surfaces seem to be a key to the attractiveness of many European residential street schemes. Not only does the change from the smooth black or gray asphalt associated with the automobile signal to drivers that they are in residential territory; it is more in character with the residential environment of brick or wooden buildings and pedestrian paths. Red, pink, yellow, ochre, or other light colors are more congruent with the colors of houses. Bricks, cobbles, gravel, and concrete blocks, sometimes set in sand (allowing grass to grow through), have all been used to slow traffic. Some routes should be kept smooth enough for bicycles and wheelchair travel.

2. *Traffic control devices and street designs should be easy to maintain.* Traffic control devices and street designs should allow for easy maintenance, and be graffiti- and vandal-proof, while encouraging residents to maintain and care for them. Designs which the residents have selected, planted, or even built for themselves are much more likely to be cared for. Mechanical maintenance techniques developed for conventional streets may have to be modified for these new kinds of residential streets. Residents may be willing to support the extra cost of manual rather than machine sweeping and snow removal (or do it themselves) in order to retain a particular residential character.

3. *The landscaping used for street design should be safe for pedestrians.* Trees or low shrubs are perceived as safer than hedges and high shrubs, because areas are more visible to the pedestrian.

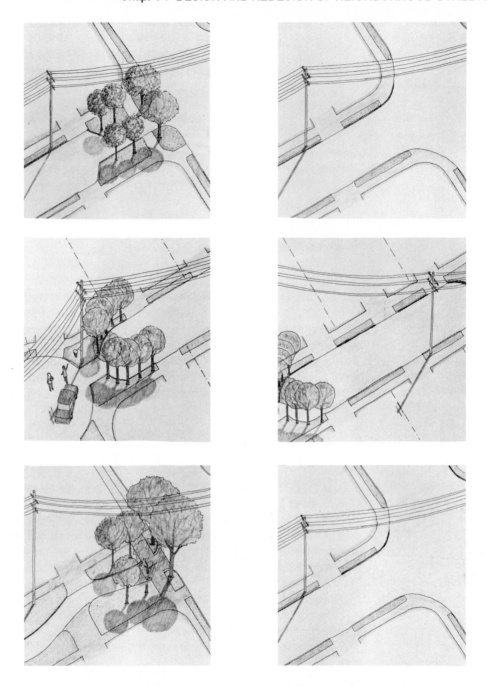

4. *Street trees should be planted to enhance the image of a street as a place with which residents can identify.* Large trees provide leafy canopies and welcome shade, screen the parked cars and the traffic, break visual continuity, soften the character of the street, and frequently enhance property values. However, they need more than a 3-foot [1-m] sidewalk strip to grow. They also require penetrable surfaces such as gravel and sand. To prevent sidewalk cracking and interference with utility lines, public works officials favor smaller "lollipop" trees. These provide little shade and tend to be petty and ornamental. They fail to impart a truly dignified character to the neighborhood. Traffic control devices can provide the space needed to plan and grow large trees. If budgets are tight, a few large trees can be interspersed among young saplings to achieve immediate effects. Immature plantings alone may not have visual impact for years.

Designers should consult relevant tree guides for the region to find trees that are tough enough to withstand air pollution, have minimal root problems, grow reasonably fast, and mature into shade trees (Friends of the Forest, 1985).

Resident participation in vegetation maintenance is essential to a successful tree-planting program in a time of high costs and low city budgets (Fig. 4.15). Planning careful designs which associate plantings with adjacent properties can give residents a sense of ownership and responsibility for their street trees.

One city department should be clearly responsible for street trees, with an adequate budget for their planting and maintenance when residents are unable to undertake this task. Wherever these principles have been applied, the results produced a more balanced situation between non-motorized and motorized residential street users.

Figure 4.15 Neighbors planting trees.

Designing for the Appropriate Travel Speed

It is an internationally accepted rule that motorized traffic must drive slowly in residential districts. An emergency stop at 35 mph [55 km/h] takes a stopping distance ranging from 115 to 165 feet [35 to 50 m] for an automobile. This distance is not acceptable so long as drivers can be faced at any moment by children or pets playing or suddenly crossing the road.

Moreover, research has shown that pedestrians are usually not seriously injured when hit by a car moving at a speed of less than 20 mph [30 km/h] at the time of impact. If impact speeds are between 20 and 35 mph [30 and 55 km/h], injuries are usually serious, while above 35 mph [55 km/h] they usually endanger life or are fatal.

Designers need to understand the relationship of their designs to travel behavior and especially to the travel speeds of drivers, pedestrian circulation, and bicyclist behavior. An activity diagram (Fig. 4.16) can show how different types of users will move through the street space. The lines or cones of sight of motorists need to be drawn. Critical viewing

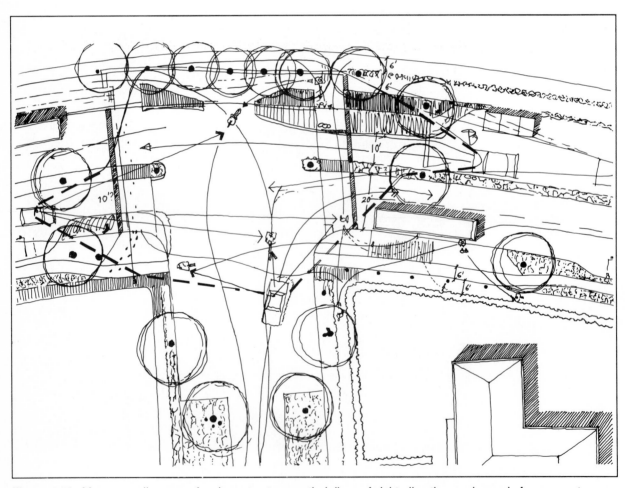

Figure 4.16 Movement diagrams of various street users, their lines of sight, direction, and speed of movement.

distances need to be explored between motorists and other users. For example, a pedestrian about to cross the road needs to be seen from a minimum distance in order for the motorist or bicyclist to react and bring the vehicle to a stop at a safe distance. In many instances it will be necessary to draw section drawings or eye-level perspectives. Scale models can be a very useful method for testing designs. Photographs taken at eye-level inside these models are a very realistic form of presentation. With special equipment such as the Simulator at the University of California, Berkeley (Bosselmann and Craik, 1986) movement through alternative street designs can be explored and recorded for evaluation before the design or plans are implemented.

Design Guidelines

Street Design Appropriate for Neighborhood Character: Not all residential streets are the same. Streets are part of the fabric of cities. Some streets serve high-density inner-city neighborhoods, others suburban or rural districts. Many streets in older cities were laid out prior to the invention of the automobile. They are often narrow and congested. Others were laid out at a time when planning for the automobile was given great importance. Here streets are frequently too wide in relation to the density they serve and to the appropriate travel speed. In each of these settings, streets with different traffic volumes exist: the street with an occasional car and mostly parked cars; the more frequently traveled street with not more than 2,000 cars per day; and the street with high traffic volumes, frequently commuter routes and traffic arteries with 10,000 or more vehicles per day.

The design for each of these streets will differ. Design should respond to traffic volumes as well as to the character of the neighborhood. However, all designs address the same problem: balance of power between the nonmotorized and the motorized users of residential streets (Appleyard and Bosselmann, 1981). On the following pages, pairs of drawings illustrate the contrast between existing conditions and possible redesigns.

Figure 4.17 Local street in an inner-city neighborhood.

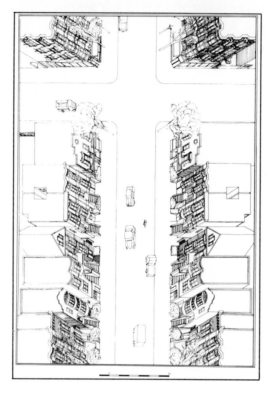

Local Streets in an Inner City Neighborhood (Fig 4.17): Residents in high-density inner-city neighborhoods frequently have less access to private open space. Street space here can be shared between the cars and the pedestrians. The space can be used to play, sit, and meet, as well as to park, drive, and deliver. A raised intersection or an adapted "Dutch Woonerf" can be an appropriate design solution. Car drivers would enter a street space where design, paving, and alignment would signal to the driver the need for greater caution. Naturally, all design changes would need to consider access for emergency vehicles and garbage trucks, as well as for delivery vans to stores, restaurants, and other commercially used properties.

In general, designs need to allow for the turning radius of fire engines and the design of paving must consider the weight of garbage trucks. However, it can be a serious design limitation if these dimensions are taken too much for granted. Small emergency and garbage collection equipment is available. Communities in hilly areas with steep and winding roads have always had a need for small trucks. If many streets in inner-city neighborhoods are converted, fire department and public works departments have the possibility of acquiring more suitable equipment.

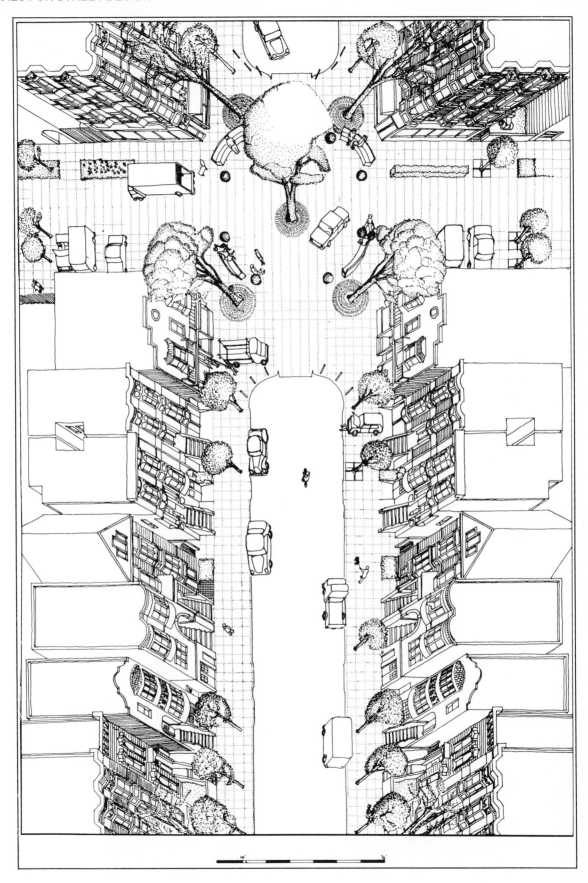

Figure 4.18 Collector street in an inner-city neighborhood.

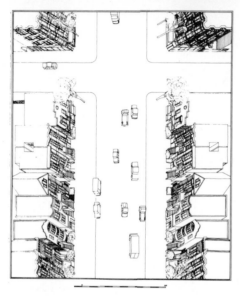

Collector or Major Street in an Inner-City Neighborhood (Fig. 4.18): Improving the livability of streets with low traffic volumes is far easier than that of streets with high traffic flows. Here, residents are concerned about the continuous noise, the visual presence, and the perceived and real dangers from moving cars. According to Appleyard's research in San Francisco, streets with high traffic house higher percentages of elderly and other less mobile population groups than streets with lower traffic volumes in the same neighborhoods. Slowing the speed of traffic is an important objective toward more livable residential streets. This can be achieved by narrowing moving lanes to a maximum of 11 feet and in many cases to 10 or 9 feet [3.35, 3.05, and 2.75 m respectively].

Routing intercity truck traffic onto nonresidential streets should be a high priority. Allowing delivery by trucks only at restricted hours also may aid in lowering traffic flows as well as reducing noise levels.

In addition, ameliorating measures can be taken such as widening sidewalks and planting rows of tall trees. The size of these trees should be in relation to the width of the street, with taller trees for wider streets.

The residents' perspective of the street will change as well. Instead of residents looking down on cars and asphalt, trees will screen the street and frame the view onto the sidewalks. If sidewalks are wide enough, there could be room for bicycles on separate lanes. There also needs to be room for low light fixtures and bus shelters.

On some streets, there might be room for a double row of trees near intersections. This will signal to the driver a visual narrowing of the street space wherever another street crosses. This can change the experience of driving by introducing a rhythm appropriate to the speed of travel. It also helps to de-emphasize the street as a long channel of movement. Care must be taken, however, to preserve good sight distances across intersection corners; trees planted near intersections must be trimmed so that no branches are below 6 feet [2m] above street level.

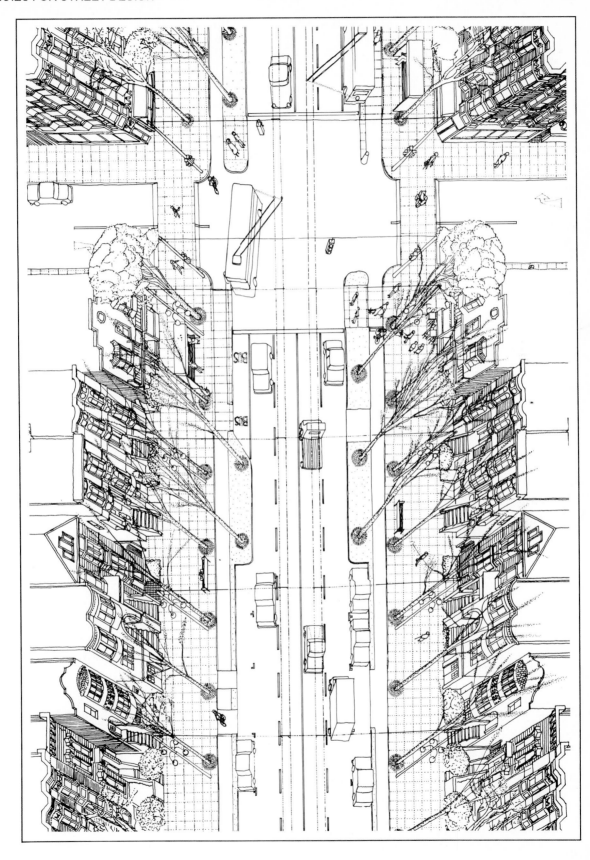

Figure 4.19   Local street in a 1920s suburb.

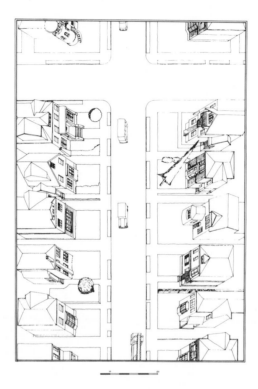

Local Street in a 1920s Suburb: Streets like that shown in Fig. 4.19 were laid out at a time when the automobile was still a novelty. Many homes have driveways with separate garage structures, frequently in the rear of the property.

On streets with light traffic, undulations can be used to slow down the occasional car driver to a pace more suitable for streets where children can play. The undulations can be designed as protected crossings and sitting areas. This makes them more visible to the driver. They can be constructed of brick and concrete, with short ramps striped for better visibility. The design of inexpensive undulations is discussed in Chapter 5. The undulation does not span the entire width of the street; this leaves enough room for bicycles to pass between the undulation and the curb. This type of undulation also allows for runoff without changing the curb drains.

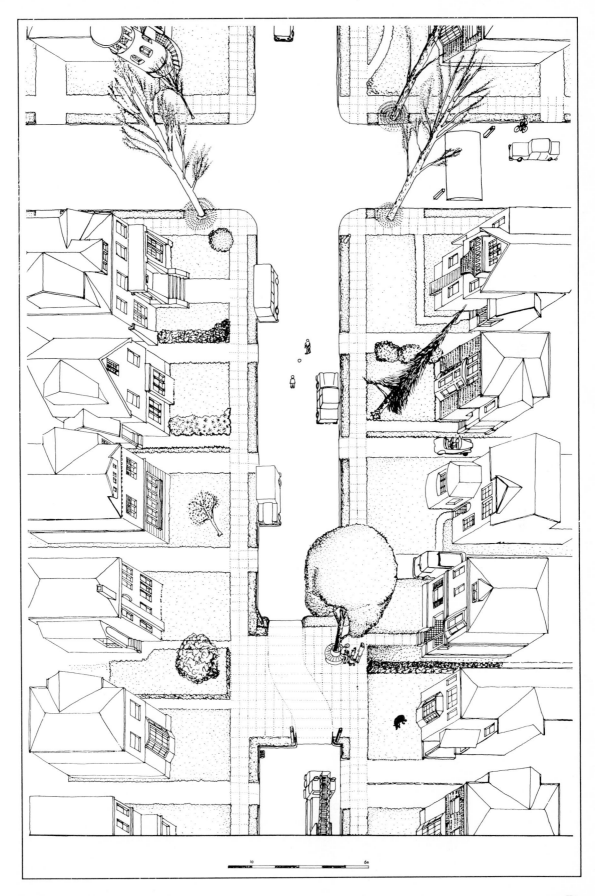

Figure 4.20 Collector and arterial street in a 1920s suburb.

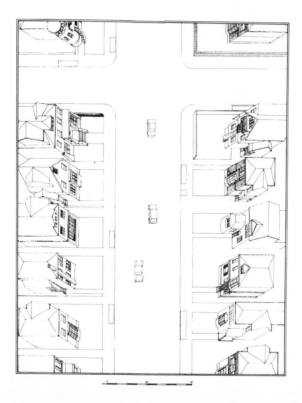

Collector or Major Street in a 1920s Suburb (Fig. 4.20): Streetcar suburbs are often close to inner cities, making the commute by public transit attractive. Bus stops need to be sheltered and should have convenient and safe crossings in their vicinity. Additional crossings might be required at midblock locations if blocks are long. They should be clearly marked through striping or pavement changes. In addition, clusters of tall trees can signal from a distance the narrowing of the street at the crosswalk. At intersections without traffic signals, the bus stop should be located on the far side of the intersection. This allows pedestrians to cross the street behind the bus.

If space permits, cycle trails should be provided on the sidewalk level, separated from the pedestrians and wide enough for cyclists to avoid the opening of car doors. Cross traffic at intersections should be slowed to allow cyclists better continuity and safety. This can be achieved by "necking down" intersections, reducing the distance for pedestrians and bicycle crossings. These neckdowns could be used for sitting areas or bus shelters.

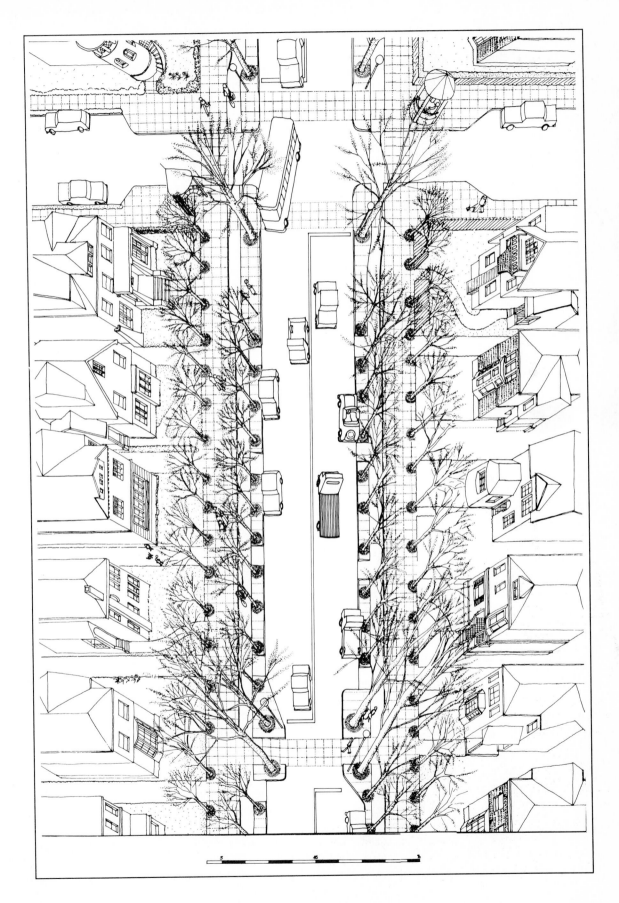

Figure 4.21 Local street in a 1950s suburb.

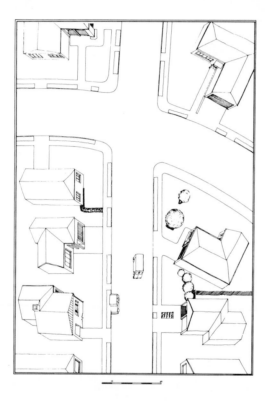

Local Street in a 1950s Suburb: Neighborhoods of the 1950s were laid out at a time when the automobile was given great importance. The width of these streets and the visibility encourage fast driving. In fact, the street shown in Fig. 4.21 is 38 feet wide (11.5m)—that is, wide enough for five cars side by side.

This street can be narrowed to a width of 16 feet (5 m) at intersections. Corners are made more rectangular, rather than curved, to discourage fast turns; however, minimum curb radii must still permit trucks and vans to make right turns (left turns where traffic travels on the left side of the road) without rear wheels climbing onto the sidewalk. Changes of direction are exaggerated by a single large tree, so as to appear more abrupt than they really are. The chokers with the trees can be designed for midblock locations, dividing long streets into identifiable shorter segments of approximately 200 feet (60 m). Between the chokers, children can play, yet be visible to drivers entering the street.

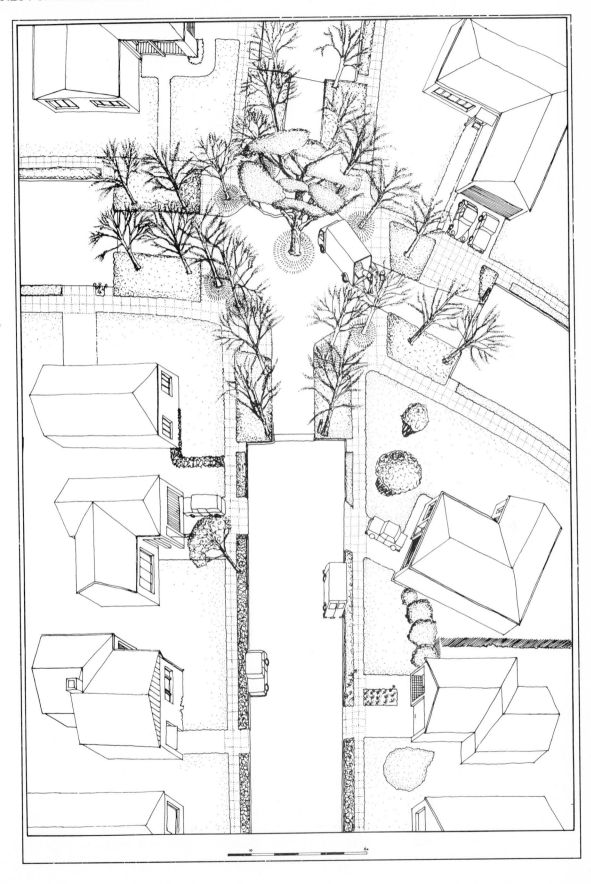

Figure 4.22 Collector street in a 1950s suburb.

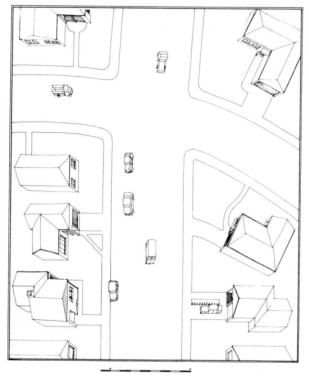

Collector or Major Street in a 1950s Suburb (Fig. 4.22): Residents in neighborhoods of "1950s suburbs" depend upon the automobile to a greater extent than do residents in the two other neighborhoods. Public transit often does not exist, or is very infrequent due to the low density. But the needs of residents in suburban neighborhoods are changing. A young generation of residents is taking over the homes from their original owners. In many young families today, both parents are working and there is a greater need for services such as child care and markets close to home. Families have fewer children, and people marry later in life. This has led to a reduction of population density. If, in 1960, 15 to 20 people lived on an acre [0.4 hectares] of land, today rarely more than 10 to 12 people do so. But the number of cars has increased with more people employed and an increasing number of unrelated adults sharing a home. In California, Connecticut, and other states, laws have been passed that allow a second unit on single-family properties in many communities. This will further increase the number of cars and trips made in and out of these neighborhoods each day. This trend will continue unless other means of transportation are made more attractive, such as neighborhood bus or van services to central transfer points located along public transit routes. Also, bicycling to transit stops and markets can be made more attractive by providing safe and interesting cycle trails plus storage at the transit stops.

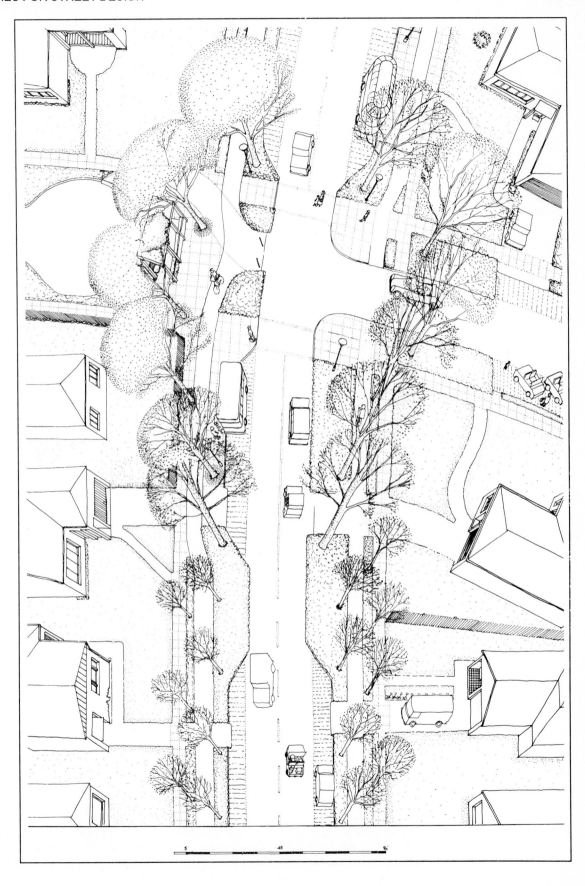

Participation. Public participation is essential for a successful street improvement plan and design. Local traffic schemes arouse powerful emotions and have widespread impact. Politically, neighborhood traffic management and street improvements are controversial because, inevitably, some people gain and others lose. A public participatory process is necessary to assess and expose potential tradeoffs before implementation. Communication with potential opposition raises the possibility of working out compromises during the planning stage. And if adverse effects are not "advertised" in advance, the fact that they do occur might be used to discredit the planning process. People are also far more likely to accept a plan or take responsibility for making it successful if they have been part of the planning and design process (Fig. 4.23).

Figure 4.23 Citizen participation in neighborhood street design.

REFERENCES

APPLEYARD, D. and BOSSELMANN, P. *Urban Design Guidelines for Street Management*. Berkeley, CA: University of California, Institute of Urban and Regional Development, 1982, working paper 385.

BERLIN, SENATOR FÜR STADTENTWICKLUNG UND UMWELT. *Verkehrsberuhigung*. Berlin, Federal German Republic, 1983.

BOSSELMANN, P. and CRAIK, K. "Perceptual Simulations of Environments." In: Bechtel, R.B., Marans, R., and Nicholson, W. (editors). *Methods of Environmental and Behavioral Research*. New York: Van Nostrand Reinhold, 1987.

GEHL, J. *Life Between Buildings* New York: Van Nostrand, 1986. (Original Danish publication with English summary: Arkitektens Forlag, Copenhagen, 1962.)

GIESLER, H. and NOLLE, A. "Akustische Verkehrsmessungen." *Zeitschrift zur Lärmbekämpfung,* v. 31, n. 2 (March 1984), pp. 31-44.

KELLER, H. *Report on Area-Wide Traffic Restraint Program*. Paper presented at the Traffic Record Systems Forum, Reno, Nevada (July 1985).

ORGANISATION FOR ECONOMIC CO-OPERATION AND DEVELOPMENT (OECD). *Traffic Safety in Residential Areas*. Paris: A report prepared by an OECD Research Group, 1979.

VISSERS, C.J. "Evaluatie-onderzoek in verblijfsgebieden, wat doen we ermee?" *Verkeerskunde 33,* (1982), n. 2, pp. 76-81.

5
Tools for Neighborhood Traffic Control

INTRODUCTION

Traffic control has as its purpose to regulate, warn, and guide vehicle operators and pedestrians in the interest of safe, efficient, and environmentally compatible movement of vehicles and pedestrians. Control may be achieved by three means:

1. *General laws and ordinances*, covering the entire jurisdiction (nation, state, city), and applying in the absence of more specific regulations, such as general urban speed limits, right-of-way rules at intersections, and general parking regulations. These controls are not discussed further in this book.
2. *Traffic control devices* that communicate specific regulatory, warning, or guiding messages to the motorist, cyclist, or pedestrian. Regulatory devices, the most common type in residential areas, include right-of-way controls at intersections, controls affecting or restricting the direction or speed of movement, and parking regulations.
3. *Geometric design features* that guide or restrict the physical movement of vehicles or pedestrians, defining and allocating various parts of the public right-of-way for use by motorized traffic, cyclists, and pedestrians, or for nontraffic uses, including landscaping.

In order to be effective, traffic control measures must be clearly understood by drivers[1] and pedestrians. To assure this, controls should convey clear, unambiguous messages; this can often be achieved by adherence to manuals which describe the various controls available and the devices used to implement them.

This chapter limits itself to the application of traffic control devices and geometric design features in residential neighborhoods and abutting streets in support of the goals and design concepts discussed in earlier chapters. For a complete discussion of traffic control methods and the devices used, the Manual on Uniform Traffic Control Devices (MUTCD) in the United States and similar manuals in other countries should be consulted. In the next two sections of this chapter, descriptions of devices and geometric design features as they have been used in neighborhood situations along with summaries of their expected effects are presented. Complete closure of street blocks is described next. The final sections discuss the effects of these controls on community services, and strategies for minimizing control violation problems.

DEVICES FOR NEIGHBORHOOD TRAFFIC CONTROL

Devices available to communicate specific controls to drivers and pedestrians include signs, pavement markings and delineators, and signals and beacons. Only traffic signals convey controls in a dynamic manner; the other devices are essentially passive, although some signs may indicate regulations or warnings which are applicable only at certain times.

This section reviews effects of those individual devices most appropriate or commonly used in residential neighborhoods. Four important devices—stop signs, speed limits, turn prohibitions, and one-way street designations—are discussed in detail, including the likely effects of their use, uniform standards and warrants for their implementation, and various related observations. Other devices are then discussed more briefly.

Two general comments in relation to the use of standards and warrants must be made. First, in some countries, including the United States, traffic engineers may be reluctant to install a control device when warrants are not met, for fear of exposing the city and the engineer to liability should an accident occur that might be attributed in whole or in part to the presence of the unwarranted device. To protect against such an eventuality, the likelihood of an accident occurring should be evaluated and, if installation still seems desirable, the reasons justifying the installation should be documented.

Second, it should be kept in mind that devices (as well as many geometric features) can be ignored or circumvented by a determined motorist. While some options are less susceptible to violation than others, any program of neighborhood traffic control will require enforcement if it is to be successful.

Stop Signs

The basic purpose of stop signs is to assign right-of-way at intersections. STOP signs are persistently requested by citizens with the expectation that they will control speed or reduce volume in residential neighborhoods. A number of studies have shown, however, that these goals are not always achieved.

In Europe, stop signs are rarely used except where sight distance limitations make less restrictive controls (such as yield signs) inappropriate.

Two-way Stop. This is used to protect traffic on one of two intersecting streets by requiring traffic on the other street to come to a complete stop. It is suitable under the following situations:

---

[1] The term "drivers" will be used to include cyclists.

— where the protected street is a major street (primarily in the United States).
— where sight distances approaching the intersection are substandard, and traffic approaching under the general rules regarding uncontrolled intersections would run a substantial risk of being involved in collisions.
— where there is a record of an accident pattern amenable to mitigation by right-of-way controls, yet conditions do not appear to justify requiring traffic on both streets to stop.

Four-way Stop. This type of intersection control is more common in the United States than elsewhere. Intended primarily where two collector or major streets intersect and where funds for a traffic signal are not available, it has frequently been used in response to complaints by the public about excessive speeds with indifferent results. The unnecessary stopping of all vehicles adds to noise, fuel consumption, and emission of air pollutants—carbon monoxide, hydrocarbons, and oxides of nitrogen.

Numerous studies have been prepared regarding the degree to which stop signs are obeyed. Generally, when not required to stop by cross traffic, only 5 to 20 percent of all drivers will come to a complete stop, 40 to 60 percent will come to a "rolling" stop below 5 mph (8 km/h), and 20 to 40 percent will pass through at higher speeds. Signs placed on major and collector streets for the purpose of speed reduction are the most flagrantly violated. Thus, stop signs which do not meet the standard warrants tend to some extent to be ignored by drivers, whereas signs placed for right-of-way purposes are more likely to be obeyed.

a. *Effect on Traffic Volume.* Where local streets offer significant savings in time over congested parallel major and collector routes or allow avoidance of congestion points, stop signs will do little to reduce traffic volume. But when the local street offers marginal travel time advantage over other routes, the time lost at stop signs may be enough to shift traffic.

b. *Effect on Traffic Speed.* Requests from citizens for installation of stop signs are usually related to desire for speed control. The general conclusion from numerous studies on effectiveness of stop signs as a speed control measure is that they have little overall effect on speed, except within approximately 200 feet (60 meters) of the intersection controlled. They are almost universally reported to have little or no effectiveness in controlling mean or 85th percentile speeds at midblock. A possible reason why resident beliefs about the speed control effectiveness of stop signs is contrary to the findings of engineering studies is that there is some evidence that stop signs do reduce the midblock speed of the *fastest* vehicles in the traffic stream. It is probably these fastest vehicles, rather than those traveling at the median or 85th percentile speed, that disturb residents. Elimination of extreme speeding by the few very fastest vehicles could satisfy the residents' concerns without altering the 85th percentile or median speeds at all.

Another reason why neighbors may feel stop signs to be an effective speed control device is that they perceive traffic slowing down and stopping at the controlled intersection as a real benefit, regardless of what effect the signs have on midblock speeds. Pedestrians are trained to cross at intersections; so a measure which reduces speeds and creates gaps in the vehicle stream there can logically be thought practical. Hence, engineering studies which have found stop signs ineffective for residential area speed control may have considered an irrelevant data base.

c. *Effect on Traffic Noise, Air Quality, and Energy Consumption.* Stop signs tend to increase noise in the vicinity of an intersection by adding acceleration and braking noise. Deceleration, idling, and acceleration increase air pollutant emissions and fuel consumption; carbon monoxide, in particular, has an adverse impact on the immediate vicinity of its emission.

d. *Effect on Traffic Safety.* The traditional traffic engineering belief is that stop signs not warranted by traffic volume conflicts or specific site safety conditions (such as inadequate sight distance) would tend to *increase* traffic accidents by inducing either a general disregard for stop signs in the community or a hazardous disregard for the specific "unwarranted" sign. Effects of unwarranted stop signs on driver behavior and safety at stop signs throughout the community are difficult to substantiate. Evidence to date on the safety effects of individual stop signs placed for volume and speed reduction purposes is mixed. It is difficult to assess reasons for these results or why the common traffic engineering belief is not more convincingly supported in the empirical data. At some of the intersections where safety decrements were measured, placement of the signs in poor visibility positions and lack of supplementary markings may account for the accident experience rather than fundamental characteristics related to the warrants. Cases where safety experience was reportedly improved may include instances where traditional warrants for stop sign installation were actually met. Further, cases which reported safety improvements may include intersections with conditions borderlining traditional warrants.

e. *Uniform Standards and Warrants.* Stop sign design details and warrants for installation are included in the MUTCD. However, the warrants relate to right-of-way assignment and response to site safety conditions; the MUTCD specifically advises that stop signs should not be used for purposes of speed control.

f. *Community Reaction.* Stop signs have a very positive image with many residents, who often see them as a solution to "near miss" as well as actual accident problems. They are also viewed as being effective at controlling speed. Negative reactions to stop signs come mainly from residents at the intersections who are subjected to additional noise from stopping and accelerating vehicles, and from motorists who think they are being stopped needlessly.

Speed Limit Signs and Speed Zoning

The speed limit sign is a regulatory device informing motorists of an absolute or prima facie speed limit imposed by the governing agency. Some signs merely remind drivers of the general limits applicable to the type of highway and area at the place at which they are posted. Where the general speed limit is not applicable to specific sites, because of special hazards, or where it is too low in view of the quality of the street (major and collector streets only) a deviation from that limit is indicated by posting appropriate signs, usually reinforced by pavement markings. The new speed limit to be imposed is determined by an engineering and traffic study of the street section involved; special attention is given to the character of the street margin—the location of sidewalks, driveways, and obstructions—and the horizontal and vertical alignment of the street, to the patterns of pedestrian use of the street, and to the existence of hazards which may not be easily detected by drivers and may therefore cause them to travel above the speed which is safe for the prevailing conditions. If no overriding safety problems are detected, the 85th percentile speed of traffic on a street is often taken as an indication of the speed zone which should be implemented.

Alternatives to speed zoning include pavement undulations and podium intersections; these will be described later.

a. *Effect on Traffic Volume.* None.

b. *Effect on Traffic Speed.* Studies evaluating the effect of speed limit signs on speed have been largely confined to major streets and high speed highways. Performance in the high speed highway cases is not considered relevant to the residential street situation. Findings in U. S. and European major surface street speed limit studies differ. In the

United States, studies have generally shown that speed limit signs have very little impact on driver speed on major streets. Drivers consistently ignore posted speed limits, and run at speeds which the drivers consider reasonable, comfortable, convenient, and safe under existing conditions. Drivers appear not to operate by the speedometer but by the conditions they meet.

In Europe speed limits have generally been effective in reducing speeds on streets, though the reductions have not always been to the limit. Studies in Great Britain, Ireland, Belgium, France, Germany, the Netherlands, and Switzerland all showed reductions in the 85th percentile speeds. Like the U.S. work, most of these studies were on collector, major, or rural highways, not local residential streets.

Explanations for the discrepancy in U.S. and European results on major/collector streets are speculative. It has been conjectured that Europeans consider conformance to a speed limit to mean traveling at or below the limit while Americans (drivers and enforcement officers) tacitly consider traveling at speeds 5 to 10 mph (8 to 16 km/h) above the limit as being compliant. It may be that American drivers rely more on their own judgment of safe and reasonable speed than on the posted limits while Europeans are more strongly influenced by the signs. Perhaps enforcement characteristics differ. Or it may be that, because speed limits are newer (not implemented until the mid 1960s to early1970s) and generally set higher in Europe than in North America, Europeans find them more reasonable, and have not yet become as cavalier about disregarding them. Moreover, narrower and more winding streets, fewer major one-way streets, and a higher proportion of "slow" vehicles may be operational reasons for closer adherence to speed limits in Europe.

Recent German studies have specifically considered the effect on residential streets of low speed limits—30 km/h (about 20 mph). Results reported seem more conformant to American experience than to prior European reports. Studies in Wiesbaden and Hamburg have found that drivers on local street do not alter their speed as a result of speed limit signs. Bremen, Karlsruhe, and Nürnberg rejected local street speed limits of this type because of ineffectiveness of the control. Berlin, Hannover, Köln, and München have introduced such local street speed limits in individual cases but traffic engineers there regard the limits' effectiveness with scepticism.

c. *Effect on Noise, Air Quality, and Energy Consumption.* If, as suggested above, speed limit signs have little or no effect on traffic speed or volume, the devices would not be expected to have any effect on noise, air quality, or energy consumption.

d. *Effect on Traffic Safety.* Excluding effects of speed limit changes on high speed highways (such as the reduction to the 55 mph limit in the United States), effects on traffic safety have been reported only in the surveys conducted in Europe. In all of those cases studied, speed zoning produced a reduction in accidents.

e. *Uniform Standards and Warrants.* Speed limit signs are a recognized control device in the MUTCD and guidelines for establishing limits are presented in basic traffic engineering references and in the laws of the various states.

f. *Community Reaction.* If speed limit signs posted are significantly lower than prevailing traffic speed, residents normally place some hope in them or in subsequent enforcement. However, if the posted limits are within a few miles per hour of the previously prevailing traffic speed, they really don't address the resident's problem. Since residents may feel that speeds of 25 to 35 mph (40 to 55 km/h) are too fast, (limits which are in force on roughly 80 percent of the residential streets in the United States), the basic issue is not whether the signs are effective but the way in which the speed limits themselves are set for local streets in the United States.

Turn Prohibition Signs

Turn prohibitions involve the use of standard "No Right Turn" (R 3-1)[2] or "No Left Turn" (R 3-2) signs, with or without peak hour limitations to prevent undesired turning movements onto residential streets. They are best used on major or collector streets at the periphery of a neighborhood to prevent traffic from entering a neighborhood altogether.

Turn prohibitions can be promulgated to be effective only during specified hours of the day, if this is desired. If shortcutting is occurring only in peak periods, restricting turns only during these periods can allow residents full accessibility during the remainder of the day.

Since turn prohibitions are clearly a passive device, their success will depend on their general acceptance by the affected drivers. In areas where regulations are frequently flaunted or poorly enforced, they will have relatively little effect. Their effectiveness may also be reduced if they seem illogical to drivers, especially when convenient alternatives to the prohibited turns are not provided.

a. *Effect on Traffic Volume.* Turn prohibition signs have been shown to have a significant effect in reducing turning volumes, though violation in the range of 10 to 15 percent of the original turning volume may be expected. The effect of turn prohibition signs is thus significant, though less than that of physical barriers. Actual traffic reduction potential depends on the percentage of total traffic on the street which the turning movement to be prohibited comprises.

b. *Effect on Traffic Speed.* To the extent that elimination of turns increases capacity on the street from which the turn is prohibited (as often would be expected), the result might be higher speeds on that street. If the movement being prohibited had formerly been used by a driver population as a speedy through route, significant reductions in speeds experienced are possible.

c. *Effect on Noise, Air Quality, and Energy Consumption.* Noise reductions are proportional to reductions in volume. Effects on air quality and energy consumption can be presumed to be negligible.

d. *Effect on Traffic Safety.* The traditional rationale for turn prohibitions has been to improve traffic flow and safety along major and collector street corridors. There is no reason to believe the device's site safety performance differs from when it is used for conventional traffic control purposes. However, as with conventional applications, there is the possibility that the prohibitions will force motorists to make turns at less safe locations or by means of hazardous maneuvers. Hence, in considering any installation of turn prohibitions, whether for conventional traffic engineering purposes or for neighborhood traffic management, the analyst should determine that safe and reasonable alternatives to the proposed prohibited movement do exist.

e. *Uniform Standards and Warrants.* Turn prohibition signs (right and left) are officially recognized MUTCD devices.

One-Way Street Designation

One-way streets can be used in several ways to protect a residential area. The traditional technique is to develop a major one-way couplet to increase capacity in a corridor; if effective, the improved operations can draw some traffic formerly using local streets onto the major streets. In a residential area, however, this technique is rarely appropriate, since

---

[2]Parenthetical references are device identification nomenclature from the *Manual on Uniform Traffic Control Devices.*

there is seldom a second, parallel major street available and since upgrading a parallel local street to major street status is usually inconsistent with the land uses along that street. Such a one-way street treatment, therefore, would simply transfer traffic impacts from one or more lightly traveled residential streets to the selected one which would probably become severely impacted. This may or may not be an acceptable choice.

Another, more successful, technique is creating a maze of one-way streets to make travel through a neighborhood difficult, if not impossible. This is done by designating selected blocks of local streets for one-way operation, making through routes difficult to find. One-way designation may also be used for very narrow streets, or to solve an inter-section capacity problem by operating the local street one-way away from such an inter-section.

The use of one-way streets has the great advantage of being a standard control that is well accepted by the public. It also provides a minimum impedance to emergency vehicles, which can travel "wrong way" when necessary. When converted to one-way operation, narrow streets where parking had been prohibited can often gain a parking lane, thus providing an added benefit to residents.

As with many nonphysical controls, one-way street systems are subject to deliberate violation, but experience shows a rather low violation rate, perhaps due to the fact that any violation will occur over a period of several seconds or minutes—whatever the time needed to traverse an entire block or blocks—whereas other devices require only a short and fast period of violation. Violation of one-way streets is more likely to be pointed out to the motorist by residents and pedestrians than are violations of other devices.

Many bicyclists can be expected to ignore one-way regulations. In the Netherlands, cyclists are permitted to travel in both directions on most local one-way streets. Signs are provided at both ends of each block, warning motorists to expect two-way bicycle traffic and indicating to cyclists that this is permitted (Fig. 5.1). On collector and major streets, a contraflow bike lane is necessary in such circumstances (Fig. 5.2).

In the United States, in contrast, cyclists generally are required to obey the one-way regulations. If two-way bike travel is desired, explicit provision should always be made by means of a contraflow bike lane (as has been provided, for example, in Eugene, OR and Madison, WI).

a. *Effect on Traffic Volume.* One-way streets used to create discontinuities in a street system have shown a high level of effectiveness in reducing through traffic.

b. *Effect on Traffic Speed.* Speeds tend to be higher on one-way streets. On major one-way couplets developed to attract traffic away from residential streets, this is a desirable result. In residential street applications, the tendency toward higher speeds can be counteracted by limiting the number of blocks with one-way continuity. Use of one-way streets to eliminate shortcuts may exclude a driver population which formerly used the streets as speedy through routes. Hence, speed reductions may be realized.

c. *Effect on Noise, Air Quality, and Energy Consumption.* On major one-way couplets, good traffic engineering practices (such as good signal progression) can mini-mize stops and starts and can thus reduce noise, emission of pollutants, and energy con-sumption. In residential street applications, to the extent that one-way maze schemes increase path lengths and necessitate slowing and stopping at turns, they may have un-desired effects on air quality and energy consumption. Noise reductions can be expected to parallel traffic volume reductions.

d. *Effect on Traffic Safety.* One-way streets tend to be inherently safer than two-way streets, because the "friction" from an opposing traffic stream has been removed. But in residential areas, where irregular patterns of one-way streets are used, careful treatment is essential at intersections where one-way blocks signed in facing directions meet and where a two-way street faces a one-way block in the opposite direction.

(a)

Figure 5.1 One-way street with signs permitting two-way travel by bicycles and mopeds.

(b)

(c)

e. *Uniform Standards and Warrants.* One-way streets are a traditional traffic engineering measure and signs and markings related to one-way operation are included in the MUTCD.

Desirable design features

- Street grid discontinuity
- Maintenance of reasonable access routes for local residents and visitors

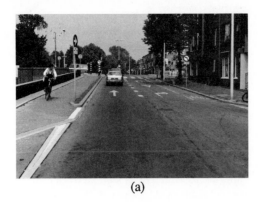
(a)

(b)

Figure 5.2   One-way collector streets with contraflow bike lane.

- Preservation of emergency vehicle access
- Minimizing of the length of one-way continuity to reduce speeding
- Use of "No Thru Traffic" signs to prevent inadvertent entry of through traffic
- Limited channelization (paint or paint and bars) at the point where opposing one-way streets meet

Undesirable features

- Generally longer trip lengths; confusing for occasional visitors; may have more stops and starts.

Other Regulatory Devices

a. *Traffic Signals.* Because of their high cost, traffic signals are used only where certain criteria of heavy traffic volumes and/or accident experience suggest this to be necessary. Signals almost always are used only where a major street intersects another major street, an important collector, or freeway ramps.

Even though confined to major streets, signals can affect local streets. Delays at signals are a prime reason for shortcutting through neighborhoods. An efficiently run major street signal system can do much to reduce through traffic diversion to neighborhood streets. Studies carried out under the Fuel Efficient Traffic SIgnal Management (FETSIM) program in California have found that retiming of signal systems for more efficient operation has reduced delays by more than 15 percent, decreased stops by 16 percent, and cut travel time by 7.2 percent.[3] Thus, travel on the major street network becomes more attractive when compared to that using local street alternatives.

Signals can also be used to improve control of access to neighborhoods by making entrance into and exit from the neighborhood easy at collector street intersections and—in the absence of signals elsewhere—difficult where local streets intersect major streets.

In some cases, the presence of a signal at the intersection of a major and a local street has made it easier for cars to get across the major street and, as a result, has the undesired effect of supporting the use of the local street by through traffic.

b. *Yield Signs.* Yield signs are used to protect traffic on one of two intersecting streets without requiring traffic on the other street to come to a complete stop. In the United

---

[3] At the same time, fuel use has dropped by about 8.6 percent.

States, this sign is used where sight distances at the intersection of two minor streets permit traffic on the controlled street to approach safely at 10 to 15 mph (15 to 25 km/hr)[4] or higher. In many countries, the sign is the standard for protecting the right of way of vehicles on an a major street.

Much of the evaluation of yield signs was performed in the 1950s when the signs were first introduced. No studies of yield signs as a neighborhood protection device are known. However, two studies have evaluated stop signs, yield signs, and "no control" in terms of their efficiency at low volume intersections. These studies, conducted in Indiana, suggest that at volumes below 200 veh/h on the minor approaches to an intersection, it is acceptable in terms of accidents, operating cost, and efficiency, to have no controls. From 200 to 800 veh/h, yield signs are as effective as stop signs in terms of accidents, and are superior in terms of energy and delay costs. Above 800 veh/h, stop signs are more effective in terms of safety performance. In all cases, stop signs are desirable if the sight distance is unacceptable according to the present standards.

c. *Access Regulation Signs.* "Do Not Enter," "Not a Thru Street," "Dead End," "Local Access Only," and "Thru Vehicles Prohibited" signs have all been used as regulatory or warning signs in various traffic situations and have potential use for neighborhood protection. Normally used to indicate the prohibited travel direction on one-way streets, and at the surface street end of freeway exit ramps, "Do Not Enter" signs have occasionally been used on residential streets in lieu of but with the same purpose as semi-diverters. "Not a Thru Street" and "Local Access Only" signs in the regulatory black-on-white format could conceivably be effective in reducing traffic volume on residential streets.

d. *Truck Restrictions.* Establishment of truck routes and use of truck route signing is a well-established practice used both for neighborhood protection and to keep trucks on streets with sufficient pavement strength to accommodate them. Regulations permitting truck travel on a street for a limited number of blocks and only for pickup and deliveries are also common.

e. *Parking Control.* Parking provisions and control can directly affect the volume of traffic on residential streets, particularly where these streets are heavily used for parking by commuters, shoppers, and other traffic attracted by nearby nonresidential destinations. Parking control may be the *only* effective traffic management device in a neighborhood if the problem traffic is comprised predominantly of outsiders who use the streets for parking. Sometimes the parking itself is considered by residents to be a primary problem. There are three basic control approaches to deal with outsider parking in neighborhoods: bans, time limits, and resident permit parking. All are effectuated by the posting of appropriate signs. Parking may also be limited by geometric design, as discussed in Chapter 4.

(1) *Bans on On-Street Parking.* Parking may be prohibited outright under certain circumstances:

— where the street is too narrow to allow parking on one or both sides
— on major streets, if the curb lane is required for through traffic during peak periods
— during hours when street sweepers or snow plows are scheduled to operate
— during early morning hours in neighborhoods where ample parking in garages and driveways is available, and where local policy favors clearing the streets of parked vehicles every night

---

[4] The lower value is specified in the MUTCD; the higher value applies in some more restrictive state and local applications.

Where residents have sufficient off-street parking to meet their own needs and those of visitors, a ban on on-street parking can be useful in keeping nonresidents from parking in the area. However, by freeing road space from parking, a wider, higher-capacity street results, and may lead to higher speeds or traffic diversion from nearby congested major streets.

(2) *Time-Limited Parking.* When long-term nonresident parking is a problem, a less drastic measure than a complete parking ban is to limit parking to one or two hours. Again, this approach is not appropriate when residents need the on-street spaces for their own use. Limited-time parking can accommodate service vehicles (repairmen, deliveries, and so on), while discouraging commuter use of the spaces.

Time limits may be counterproductive if the nonresident parkers are short-term users of the spaces, as in the case of shoppers, students, and business clients. Establishing time limits in these circumstances would tend to increase the effective parking capacity for such users and thus might increase nonresident traffic on the street. However, if the problem is due to special events, such as cultural or sports events, parking limits of one or two hours can reduce parking by persons attending such gatherings.

(3) *Resident Permit Parking.* Resident permit parking (RPP) is used in a growing number of North American cities to reduce nonresident parking on residential streets. RPP is most commonly applied when residents require on-street space for parking their own vehicles, but it is sometimes used simply to keep nonresidents out of the area.

There are numerous variations of RPP. Some versions bar any nonresident use of on-street parking space without a special permit (for example, Cambridge, MA), while others allow time-limited parking by nonresidents (such as San Francisco and Calgary). While the latter variation is easier for visitors of residents, it may not be a solution in cases where a large component of the nonresident parking problem is due to short-term users.

Some cities have begun to experiment with special nonresident permits for parking in RPP zones. For example, Berkeley, CA permits carpools registered through its ridesharing office to purchase such permits if pool members work in the zone.

Regardless of the type of parking control, it is effective only if regularly enforced. However, many communities have found that enforcement at a level to effectively free up parking spaces can pay for itself if fines for violations are high enough.

Guide and Warning Controls

The main guidance required in residential neighborhoods is the identification of streets. Since driving speeds are low, especially at intersections, street name signs need not be large, but they should be located in such a manner as to be readily found, especially at night.

Warning controls in residential neighborhoods also have limited uses. Usually, drivers and pedestrians need to be warned only of special hazards. The attention of drivers is drawn to the location of schools (especially elementary schools) and playgrounds, to pavement undulations, to the fact that a street may be a cul-de-sac or that traffic barriers or diverters are located ahead, and to stop and yield signs ahead if they are not readily visible because of curves or shrubbery. Warnings are conveyed by standard signs listed in manuals of traffic control devices. In hilly terrain, it may also be necessary to identify hazardous grades, sharp curves, and blind intersections.

Guide and warning control devices are frequently used to supplement geometric features created for neighborhood traffic control purposes. Some of the warning control devices most commonly used are as follows:

a. *School Zone Signs and Beacons.* "School" signs and flashing beacons are frequent and standard throughout the United States to warn drivers of the presence of school children in the area. The use of signs and flashers time-coordinated with the presence of children is believed important, since signs continuously present are not always effective. Yet, studies in Madison, WI, report that school signs without beacons are not necessarily less effective than those with beacons. The MUTCD addresses the use of these devices.

b. *Slow Signs.* While not in the MUTCD, slow signs are a frequently used warning sign at locations thought to present some hazard. Some studies suggest that such signs would have value on residential streets. However, they are vague and unenforceable; they do not convey a clear message to drivers, who are unlikely to change speed unless they see some reason other than the sign itself to do so. The result may, therefore, have little more than a placebo effect on residents.

c. *Lateral Bar Pavement Markings.* This device is comprised of bars painted laterally across a roadway half-width with decreasing spacing between the bars to give the driver either the illusion of speeding up (in the hope of inducing a conscious effort to slow down) or the sense of approach to some important road feature. Applications have been limited and this measure is regarded as an experimental device.

d. *Crosswalk Pavement Markings.* Studies indicate that crosswalks of the common and zebra design both are effective in attracting pedestrians, but the driver reaction and accident experience are not usually enhanced in comparison to unmarked crosswalks in similar circumstances. Many communities in California are removing crosswalk markings because they believe that the markings give pedestrians an unjustified sense of security when crossing. In a study in San Diego it was found that pedestrian-vehicular accidents actually were more likely in marked crosswalks. Crosswalk design is addressed in the MUTCD.

e. *Lane Reduction.* Restriping roadways to reduce the number of lanes from two to one in each direction has met with some success in Calgary.

f. *Novelty Signs.* Many attempts to attract the driver's attention through the use of unique and unusual signs have been made. Some examples include messages warning of children at play, of domestic animals crossing, of special speed-limit enforcement, and odd-value advisory safe speed signs.

Unfortunately, the novelty effect wears off quickly and the signs no longer attract the attention of regular passers-by. They are a target for vandals and souvenir hunters and have a high replacement cost. Unique message signs have no legal meaning or established precedent for use in basic traffic engineering references; their use is discouraged because of both the lack of proven effectiveness and undesirable liability exposure.

GEOMETRIC FEATURES FOR NEIGHBORHOOD
TRAFFIC CONTROL

Geometric features of the road used for neighborhood street traffic fall into three categories:

— Features which physically restrict and prevent vehicle movement; these include chokers, traffic circles, median barriers, semidiverters, forced-turn channelization, diagonal diverters, and cul-de-sacs.

— Features which physically reduce speed, such as pavement undulations and raised intersections.

— Features which attract the special attention of drivers, such as rumble strips.

Their common characteristic is that by their physical form they force or prohibit a specific action. Geometric features have the advantages of being largely self-enforcing and of creating a visual impression, real or imagined, that a street is not intended for through traffic. The disadvantages relative to other devices are their cost, the potential negative impact on emergency and service vehicles, and the imposition of inconvenient access on some parts of a neighborhood. They also are static and must be appropriate at all hours of the day and night.

Chokers, traffic circles, median barriers, diverters, channelization, and cul-de-sacs, as well as some less commonly used geometric features, are described in detail in this section. All such features are normally deployed in conjunction with traffic control devices such as signs, pavement markings, and reflectors, to warn motorists of their presence and indicate appropriate behavior.

Chokers

A choker or curb bulb is a narrowing of a street, either at an intersection (Fig. 5.3) or midblock, in order to reduce the width of the traveled way. While the term usually is applied to a design which widens a sidewalk at the point of crossing, it also includes the use of islands which force traffic toward the curb while reducing the roadway width.

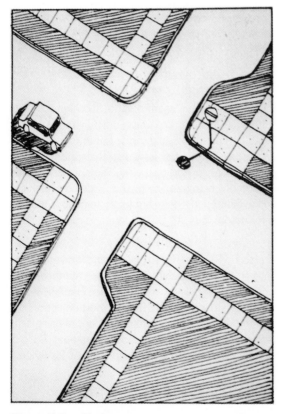
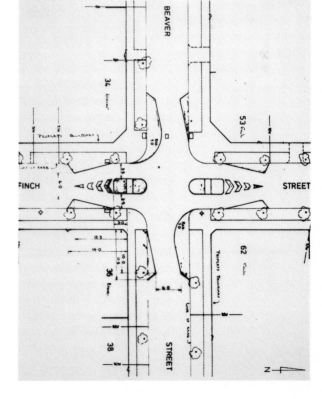

Figure 5.3   Choker.

Streets narrowed at the crosswalk reduce the distance over which pedestrians are exposed to vehicular traffic. Bulbs provide safe areas for people to walk or play, or may provide added area for landscape or gateway features, thereby improving the appearance of the neighborhood.

a. *Effects on Traffic Volume.* Studies to date have shown that curb bulbs reduce traffic volume only when they either reduce the number of lanes of travel or add friction to a considerable length of street.

b. *Effects on Speed.* Curb bulbs appear to have insignificant effect on speed.

c. *Effects on Noise, Air Quality, and Energy Conservation.* No significant effects have been identified.

d. *Effects on Traffic Safety.* Curb bulbs can improve the safety of an intersection by providing pedestrians and drivers with an improved view of one another. They also reduce pedestrian crossing distance, thereby lowering their exposure time to vehicles.

e. *Uniform Standards.* Chokers or curb bulbs can be considered to be either normal extensions of the existing curb or channelizing islands as defined in the MUTCD and parallel design manuals.

Traffic Circles

These geometric design features, also called *rotaries* or *roundabouts*, have several different functions. Large circles can replace intersections, changing direct conflicts of traffic streams into weaving maneuvers; they are used mainly in Europe and the eastern United States to increase capacity. In Great Britain, miniroundabouts are used for traffic control on minor streets. However, the circle—typically a painted dot as small as 3 to 4 feet. (1-1.30 m) in diameter—is used to change right-of-way priorities at fairly busy intersections rather than to reduce traffic volume or speed.

Circles of an intermediate size (of the order of 10 feet or 2.5 meters in diameter), the subject of this discussion, are being tried mainly as speed control devices within the intersections of two local streets; a secondary objective is to reduce traffic volumes by using them as part of a group of circles or other devices that slow or bar a driver's path (Fig. 5.4).

a. *Effect on Traffic Volume.* Studies which have examined traffic volume effects of traffic circles have also included other devices in their proximity; traffic volume effects of circles are not attributable or quantifiable to the individual circle but to the system of controls within which they are deployed. The assumption is, however, that volume reductions result from psychological rather than physical impacts on traffic. Their presence when viewed from a distance gives an impression of obstruction to traffic. If drivers have encountered real barriers at other points in the community, they are likely to believe that the circle is yet another one and change routes before they get close enough to see what it actually is.

b. *Effect on Traffic Speed.* The effect on vehicle speed has been shown to be related to the size of the circle, the distance from the circle at which speeds are measured, and the presence or absence of additional obstructions at the intersections. Overall, the effect of circles on speeds has varied: some only slow down the fastest, most objectionable vehicles and only in their immediate vicinity; others cause substantial drops in the speeds of all vehicles at the intersection. In Australia, by contrast, data show substantial drops in speed by all vehicles on roads about 7 meters (22 ft) wide where circles of less than 7 meters diameter are provided.

c. *Effect on Noise, Air Quality, and Energy Consumption.* Effects in the areas are marginal as they relate to small effective changes in speed and volume.

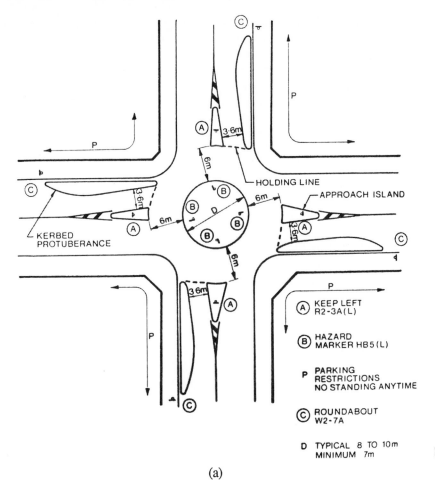

HOLDING LINE

APPROACH ISLAND

KERBED
PROTUBERANCE

Ⓐ KEEP LEFT
R2-3A(L)

Ⓑ HAZARD
MARKER HB5(L)

P PARKING
RESTRICTIONS
NO STANDING ANYTIME

Ⓒ ROUNDABOUT
W2-7A

D TYPICAL 8 TO 10m
MINIMUM 7m

(a)

Figure 5.4 Traffic circle.

(b)

(c)

d. *Effect on Traffic Safety.* Numerous observations have been made of unsafe practices caused by circles. They present an increased hazard to pedestrians by bringing vehicles, some at relatively high speeds, nearer to the curb. The deflection they cause to an automobile can also impinge upon a bicyclist's path. Observations have also been made of vehicles striking curbs or jumping over them onto lawns when diverted by circles. Vehicles have frequently been observed passing to the left of a circle when completing a left turn. Each of these is an unsafe action which can be directly attributed to the circle itself. The lack of substantiating accident statistics tend to speak more to the short time of use and use on low volume streets rather than necessarily indicating inherent safety of the circles.

e. *Uniform Standards.* Traffic circles are not a specific entry in the MUTCD. However, they may be considered to be channelizing islands which are in common use to control traffic.

f. *Community Reaction.* Community reaction to traffic circles has been mixed. Some people, particularly those in the immediate vicinity of a circle, perceive a reduction in the speed of traffic. Others perceive them to have no effect or to act mainly as a nuisance.

Desirable design features.

- *Location.* Traffic circles should not be located where pedestrian or bicycle volumes may create conflicts, as noted above.
- *Visibility.* The circle itself should be made of materials with a high target value for both day and nighttime visibility. "Keep Right" (R4-7) signs should be visible on all approaches.
- *Delineation.* Centerlines should be used on each approach to guide traffic around the circle.
- *Safety.* Crosswalks should be located out of the influence zone of the circle. Parking restrictions should be placed adjacent to the intersection to minimize conflicts with parked vehicles.
- *Size.* The circles should be large enough to impact speed, but they must permit trucks and fire engines to make all necessary turning movements.

Median Barriers

The median barrier is a standard traffic engineering installation generally used to improve flow on a major street. In the context of neighborhood traffic management, it is used at the intersection of a major and a minor street to make all left turns and the through movements on the minor street impossible (Fig. 5.5). A median island is constructed across the intersection on the major street; if the major street has a midblock median, this median is extended through the intersection. Left turns can then be concentrated at places where they can be better controlled, often with turn pockets and signals.

The median barrier is one of the few control techniques which can aid major street flow and enhance neighborhood protection at the same time. By restricting the movements mentioned above, the barrier can be as effective as a full or partial barrier or diverter in reducing traffic on residential streets. Since the median barrier is an accepted arterial treatment, it is less likely to arouse opposition than some other physical treatments.

A median barrier is most effective if applied at all local street intersections along the major street; otherwise, the effect may be to merely shift traffic from one local street to another.

Figure 5.5 Median barrier.

a. *Effect on Traffic Volume.* The use of the median barrier as a protection device has been best documented in Gothenberg, Sweden. Median barriers were used on a loop road around the central business district, resulting in a traffic reduction of 70 percent on streets inside the loop and an increase of 25 percent on the circumferential street.

b. *Effect on Traffic Speed.* Median barriers which reduce accessibility to neighborhood streets may exclude a driver population which formerly used the streets as speedy shortcuts. In this sense they might substantially change speeds experienced along residential streets.

Median barriers have been infrequently used to control speeds on small radius curves on major and residential streets. By preventing traffic on the outside of the curve from crossing the centerline to "straighten out the curve," the median barrier emphasizes the degree of curvature and causes traffic to slow.

c. *Effect on Noise, Air Quality, and Energy Conservation.* To the extent that they reduce traffic volume of speed, median barriers are also likely to reduce noise. The use of median barriers has a marginally positive effect on air quality and energy conservation when they improve the quality of flow along a major street. Some of these benefits can be lost, however, if turning movements become so concentrated at specific locations that excessive delay and waiting time occurs to turning vehicles. On the other hand, it may be more cost-effective to invest in special turning lanes, turn phases at signals, and so forth at a few intersections than to maintain free turns at many.

d. *Effect on Traffic Safety.* Studies of median barriers have shown that they improve the safety of the major street, and that the improvement is inversely proportional to the number of openings permitted in the median. The effect of safety of local streets has not been quantified, but a reduction in accidents proportional to reductions in traffic can be presumed.

e. *Uniform Standards.* Median barriers are a recognized MUTCD device and are provided for in state design manuals.

Desirable design features

Details of effective design for median barriers are contained in many design manuals. Warrants and design details are also found in NCHRP Report 93, "Guidelines for Median and Marginal Access Control on Major Roadways," and NCHRP Report 118, "Location,

Selection, and Maintenance of Highway Traffic Barriers." These publications are concerned primarily with the barriers' effect on major streets. Desirable design features in regard to protection of local streets include:

- *Location.* Location of a median barrier should include consideration of prevention of through traffic and shortcutting; however, access to major traffic generators and emergency facilities must be maintained.
- *Visibility.* Visibility is rarely a problem since major streets with medians are usually well lit. Use of reflectorized buttons and "Right Turn Only" (R 3-5) signs can improve visibility of the median from the local street.
- *Safety.* The end point of median island should be designed to minimize damage to a vehicle that strikes the end of the barrier. If pedestrians are permitted to cross the major street, protection can be accomplished by use of sufficiently wide medians to give the pedestrians "trapped" on the island a relative feeling of safety.
- *Emergency Passage.* Emergency passage across a median barrier is provided at regular median openings. In addition, mountable curbs or ramps and paved emergency vehicle passage can be provided.

Depending on the width of the median, it may provide opportunities for urban gardens or other special neighborhood-oriented landscape treatments, or even for recreational uses. The very wide median along Commonwealth Avenue in Boston, MA, is famous for its public art, playgrounds, and garden plots.

Semidiverters (Half-Closures)

A semidiverter is a barrier to traffic in one direction of a street which permits traffic in the opposite direction to pass through. In a sense, it is a physical reinforcement to a regulatory "Do Not Enter" sign and is normally accompanied by such a device, as well as by turn prohibition signs on the crossing street (Fig. 5.6). It is an alternative to using a one-way street designation for the same block, and allows residents on the block limited two-way travel opportunity.

Figure 5.6   Semidiverter.

Because they block only half of a street, half-closures are easily violated, particularly on low volume streets. Their advantages over full barriers or cul-de-sacs are that they provide a minimal impediment to emergency vehicles and cause less interference to local traffic. Experience has shown that they work best in areas where neighborhood traffic management is generally well accepted by the public.

a. *Effect on Traffic Volume.* Half-closures can make significant reductions in volume, though residents may often focus on the violation level rather than the reduction level. Traffic reductions of 40 percent of the prior two-way volume are common, implying violation rates of 10 percent of the former volume.

b. *Effect on Traffic Speed.* A semidiverter does not reduce speed per se. However, if it diverts drivers who formerly used the street as a speedy through route or shortcut, the actual change in speed experienced after installation may be substantial.

c. *Effect on Noise, Air Quality, and Energy Consumption.* Effects on noise levels are directly related to the reduction in traffic volume. As with most devices considered here, the air quality changes in the micro-environment are minuscule since most auto-related pollutants which affect neighborhoods are responsive to changes in emissions on a regional basis rather than that in a small, localized area. Energy consumption can be assumed to be somewhat increased due to slightly longer distances and added stops on major streets.

d. *Effect on Traffic Safety.* Half-closures appear to shift locations where accidents occur to other streets rather than reducing overall accident experience.

e. *Uniform Standards.* Half-closures are not included in the Manual of Uniform Traffic Control Devices. Like other geometric features they define an area which is not in the traveled way and can be comprised of elements included in the MUTCD and other design manuals. Half-closures are recognized as residential traffic control treatments in some basic traffic engineering reference texts.

f. *Community Reactions.* People living on streets with a half-closure have been generally favorable to them. The major negative reactions have been due to the observed violations and lack of enforcement to prevent them.

Desirable design features

- *Location.* A half-closure is best located at the end of a block to prevent entrance and allow exit. Half-closures located to prevent exit are easily and frequently violated. Half-closures in midblock locations are also more frequently violated than end-of-block placements, though they still have some effectiveness.
- *Visibility.* "Do Not Enter" (W 5-1) signs, painted curbs, and reflectorized signs and construction materials are useful for aiding visibility.
- *Violation Prevention.* Constriction of the traveled way in the direction in which traffic is permitted to pass the diverter can make violations more difficult.
- *Emergency Passage.* This is inherently permitted by the design of the device. As long as sight distance is good, it is quite acceptable for emergency vehicles to travel in either direction on the "open" side of the half-closure. However, if traffic is queued on the "open" side, awaiting a gap in cross-street traffic, emergency vehicle passage can be delayed.
- *Bicycles and the Handicapped.* Care should be taken to provide a legal bypass for bicycles and wheelchairs; otherwise cyclists, in particular, will violate the device by riding "wrong way" on the open side, endangering themselves in the process.

Forced-Turn Channelization

Forced-turn channelization usually takes the form of traffic islands specifically designed to prevent through traffic from executing specific movements at an intersection. Its basic function is to make travel on local streets difficult, but not prevent it entirely. Generally this technique is best used at an intersection of a major and a local street, where the major street is basically unaffected by the channelization, or even has its traffic flow qualities enhanced, while through traffic on the local street is prevented. Employed in such locations, it prevents traffic flow from one neighborhood to another across the major street. It can also be used on purely local streets to permit turning movements other than those possible with a diagonal diverter. However, it is more likely to be violated within a neighborhood, since the threat of enforcement is minimal.

Forced-turn channelization can take numerous forms and must usually be customized to deal with specific traffic movements to be prevented. Because channelization has become well accepted and well used in traffic control, it tends to have a higher level of obedience than the other partial restrictions, particularly when used on a major street.

  a. *Effect on Traffic Volume.* Channelization is effective in reducing local street volume if the movements prevented are significant contributors to overall traffic on the street.

  b. *Effect on Traffic Speed.* Channelization tends to have a minimal direct effect on speed, other than the required slowing for turning. But if it diverts a driver population which had previously used the street as a high speed through route, the actual change in speeds experienced on the street may be marked.

  c. *Effect on Noise, Air Quality, and Energy Consumption.* Noise reduction is likely to parallel the reduction in volume and speed on the local street. While this technique does add some slowing and accelerating, as well as added distance on other routes, air quality and fuel consumption effects are likely to be negligible.

  d. *Effect on Traffic Safety.* Channelization tends to increase the safety of locations where the design adds clarity and simplicity and is easily understood.

  e. *Uniform Standards.* Channelization islands are covered in the MUTCD, which recognizes their use to control turning movements.

  f. *Community Reaction.* Programs involving channelization have generally been acceptable to communities where proper planning and communication have occurred. Specific problems have come from specific individuals or high volume traffic generators whose access has been impaired. Citizens have also complained where the design did not adequately prevent violations from occurring.

Desirable design features

General practices in the area of channelization including design of effective turn radii, merging distances and visual clarity, are applicable to this device. Other desirable features include:

- *Visibility.* Channelization will usually be constructed of some type of raised material, either curb and gutter, concrete bars, or asphalt berm. Painting the devices white will add to the visibility, as will standard signs (such as "No Right Turn," (R 3-1) and "No Left Turn," (R 3-2)) indicating the turns permitted and/or prohibited.
- *Delineation.* Striping parallel to the device and continued (dashed) through the intersection aids in the clarity of the design.

- *Violations*. Prevention is best done by assuring that the channelization covers a significant part of the intersection, thereby narrowing the area where illegal turning movements can be made.
- *Emergency Passage*. High speed emergency passage is generally difficult to provide for without also providing for easy violation of the intent of the device. However, emergency vehicles can usually maneuver around channelization without severe delay.
- *Pedestrians, Bicycles, and the Handicapped*. Special care should be given to providing routes for bicycles through a channelized area; otherwise, cyclists will tend to make their own way, often in violation of the channelization and sometimes at a hazard to themselves. Islands designed to give adequate refuge for pedestrians should also provide for ramps for wheelchairs.

Diagonal Diverters

A diagonal diverter is a barrier placed diagonally across an intersection to, in effect, convert the intersection into two unconnected streets, each making a sharp turn (Fig. 5.7). The primary purpose of a diagonal diverter is the same as that of forced-turn channelization—to break up through routes, making travel through a neighborhood difficult, while not actually preventing it.

If used at a single site, the diagonal diverter is effective only when the neighborhood it is intended to protect is a limited one. If the neighborhood is larger, with other continuous residential streets parallel to the "problem street," installation of a single diagonal diverter may merely shift through traffic to another local street rather than to bounding major and collector streets. In actual application, this device is therefore often best used as part of a system of devices which discourage or preclude travel through a neighborhood.

The basic advantage of a diagonal diverter over a cul-de-sac is that, by not totally prohibiting the passage of traffic, it tends to reduce the circuity of travel imposed on local

(a)

(b)

Figure 5.7 Diagonal diverter.

(c)

(d)

residents and maintains continuous routing opportunities for service and delivery vehicles. It also does not "trap" an emergency vehicle and can be designed to permit the passage of some emergency vehicles through the diverter.

a. *Effect on Traffic Volume.* Studies of systems of diverters have found that traffic on streets with diverters can be reduced from 20 to 70 percent depending on the system of devices in the area. In these studies, traffic on nondiverted streets increased by as much as 20 percent. Obviously, the amount of traffic reduction in any case is highly dependent on the amount of through traffic at the problem site originally.

In general, a pattern of devices that turns a neighborhood with a gridiron street pattern into a maze tends to be successful whereas systems which use diverters together with devices such as traffic circles or stop signs are less successful. Diagonal diverters tend to be less successful in areas where heavy pressure on surrounding major streets makes the diverter-created maze of local streets still preferred by some drivers.

b. *Effect on Traffic Speed.* General comments from citizens and agency staff suggests that diverters are most effective as speed control devices only in their immediate vicinity, within about 200 to 300 feet (about 60 to 90 m) of the device. However, they are not primarily installed for the purpose of speed control. On the other hand, diverters may eliminate from the street a driver population which formerly had used it as a speedy through route. As a result, the effect on speeds experienced along the street may be substantial.

c. *Effect on Noise, Air Quality, and Energy Consumption.* No effects, other than through shifts in volume of traffic and changes in speeds, are to be expected.

d. *Effect on Traffic Safety.* Evaluations of accident experience before and after the installation of diverters show a significant reduction in the number of accidents in the neighborhood; however, the actual number in each case was quite small. Accidents were shifted to major streets, which picked up traffic volume (note that on major streets a more effective program for minimizing accidents may be possible).

e. *Uniform Standards.* Diagonal diverters are not specifically listed in the MUTCD or parallel state traffic control and design manuals. However, they may be defined as a channeling island and may be constructed of and marked with devices shown in the MUTCD or other design manuals. Diverters are specifically recognized as a form of traffic control channelization in basic traffic engineering texts.

f. *Community Reaction.* Appleyard's studies in the Clinton Park neighborhood of Oakland, CA, found residents on streets with diverters substantially more satisfied with their neighborhood than residents on other streets. Reaction revolves around whether specific individuals feel they benefit or lose as a result of the installation. Residents generally tolerate the slight inconvenience in access which a diverter creates; they are less tolerant when it adds traffic to their street. They are more likely to criticize diverters in neighborhoods other than their own, since they perceive those located distant from their home from a driver's rather than resident's perspective. Voting patterns for two initiatives to remove diverters in Berkeley substantiate this: in areas of the city where few diverters are located, a majority of voters voted for removal; areas of the city where most of the diverters are located voted overwhelmingly to retain them.

Desirable design features

For a diagonal diverter, the following features should be incorporated for a safe and effective design:

- *Visibility.* The device should be easily visible both during the day and night. Features such as painted curbs and rails, button reflectors, black and yellow directional arrow signs (Sign W1-6), street lighting, and elevated landscaping can produce a highly visible diverter.

- *Delineation.* Centerline pavement striping supplemented in those areas where weather permits by pavement buttons are useful in further identifying the proper driving path.
- *Safety.* In addition to the visibility items, materials that do relatively little damage if hit are desirable. These would include shrubs rather than trees for landscaping; breakaway sign poles; and mountable curbs if the device contains additional material besides the curb to prevent violations. Temporary barricades of wood and asphalt berm can also be both safe and effective.
- *Emergency Passage.* Designs which permit emergency vehicles to pass while restricting auto passage are desirable. Since an open emergency vehicle gap is an invitation for other vehicles to pass through as well, some physical control of the passage may be necessary. Designs employing gates or chains require emergency vehicles to either stop and open the gate or to "crash" through it. Enforcement should insure that parked vehicles do not block the emergency passage.
- *Violation Prevention.* Clearly, the device itself should prevent normal traffic from passing through. Additional wooden or steel posts along the adjacent properties may be needed to prevent drivers from driving on lawns, driveways, or sidewalks in deliberate violation of the device.
- *Drainage.* A design which stops short of the existing curbline will usually allow existing drainage patterns to be maintained, thereby reducing overall costs.
- *Pedestrians, Bicycles, and the Handicapped.* Provision should be made for the continuity of bicycle routes around the diverter. Use of sidewalk ramps for bicyclists can also aid persons in wheelchairs. Extension of sidewalks across the diagonal can provide a safe pedestrian crossing.
- *Signage.* It is usually desirable to supplement the physical control offered by the diverter with regulatory signs, such as "right (left) turn only."

Cul-de-Sacs at Intersections

An intersection cul-de-sac is a complete barrier of a street at an intersection, leaving the block open to local traffic at one end, but physically barring the other (Fig. 5.8). As such, a cul-de-sac represents the most extreme technique for deterring traffic short of barring *all* traffic from the street in question.

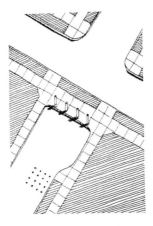 

Figure 5.8   Cul-de-sac at intersection.

A cul-de-sac created from what was a through street can still allow emergency vehicle passage, such as at a hospital in Berkeley, CA, where a park was created using "grass-crete" (concrete matrix with grass planted in cells), laid out to permit emergency vehicle passage.

Since a cul-de-sac is completely effective at its task of preventing through traffic, the choice of where and whether or not to use it depends largely on other aspects of traffic movement. For example, a cul-de-sac is less desirable in the vicinity of fire, police, or ambulance stations where emergency vehicle movements are frequent. It is less desirable than other devices in areas where the potential for multi-alarm fires might exist, since fire departments often wish to maximize the flexibility of vehicular movement in these places.

In locations near major traffic generators, a full barrier may be the only method of preventing shortcutting. On the other hand, the design of the cul-de-sac must often allow side or rear access from a local residential street to a high traffic volume generator fronting on a major street; in this case, a midblock cul-de-sac, discussed in the following section, may be more appropriate. A cul-de-sac may be desirable adjacent to a park or school where the vacated street can be converted into additional play space. Finally, a cul-de-sac may be considered as a last resort in locations where obstinate drivers violate other less effective devices.

a. *Effect on Traffic Volume.* Cul-de-sacs are extremely effective at limiting traffic volume and normally reduce traffic to that generated in the immediate local vicinity. Exceptions are the occasional vehicle which unknowingly enters a blocked street and then must maneuver to leave it, and those few vehicles which deliberately violate the barrier. Signing described below should be used to reduce such incidents.

b. *Effect on Speed.* A cul-de-sac is not a speed attenuating device. However, if the device eliminates a driver population which previously had used the street as a speedy through route, its ultimate effects on traffic speeds experienced on the street may be substantial.

c. *Effect on Noise, Air Quality, and Energy Consumption.* Noise has been found to be reduced as a function of the reduction in traffic volume and speed. Air quality and energy consumption effects are negligible.

d. *Effect on Traffic Safety.* Safety effects of cul-de-sacs are similar to those noted for diagonal diverters.

e. *Uniform Standards.* Permanent cul-de-sacs are a standard treatment in the design of new residential developments. Basic traffic engineering reference texts acknowledge the use of retrofit cul-de-sacs for residential traffic management.

f. *Community Reactions.* Communities have generally responded positively to cul-de-sacs, particularly where a number of such treatments have been installed in a neighborhood. They have been less well received where they merely shift traffic from one street to another. Some resentment occurs if a long detour for access is caused by a series of barriers.

g. *Effect on Emergency and Service Vehicle Access.* The cul-de-sac or complete barrier of a street is the neighborhood protective device most objectionable to emergency and service personnel. While traversable barriers (see the Section on Traversable Barriers, page 107) can accommodate emergency vehicles, even these can be rendered ineffective by cars parked in front of the opening. More so than a diagonal diverter, a complete barrier can cause considerable interference in the proper placement of vehicles combating a fire.

Police vehicles giving chase to a pedestrian suspect can occasionally be inhibited by a cul-de-sac with or without an emergency vehicle passage. However, some cul-de-sac designs will stop the suspect as effectively as they will stop the police vehicle.

Desirable design features

A successful cul-de-sac design should incorporate the following features:

- *Location.* Where possible, the barrier forming the cul-de-sac should be placed at an intersecting through street rather than in the interior of a neighborhood. Location in this manner minimizes the inadvertent entrance into the closed street and subsequent maneuvering to exit. Signing can help alleviate this problem when midblock location of the device is desired.

  If adequate turning radii, as specified in the next paragraph, cannot be provided, cul-de-sacs should not be employed on relatively long blocks.

- *Turning Radius.* A typical minimum turning radius standard for cul-de-sacs in new subdivisions is 35 feet (10.5 m). This permits free 180° turning movements by autos and smaller trucks and maneuvering space for larger vehicles. When an existing residential street, perhaps only 36 to 40 feet (11-12 m) wide is made a cul-de-sac, only an 18 to 20 foot (5.5-6 m) turning radius can be provided in the existing traveled way. Bulbing the turning area beyond the existing curblines should be considered where feasible. Parking bans on the approaches to the turning area can also help ease turning movements, but must be enforced to be effective. Enforcement, however, may be costly to provide in an area otherwise not receiving regular patrols from parking enforcement staff.

  The need for an adequate turning radius is greatest at cul-de-sacs located within a neighborhood where inadvertent entrances to a block are more likely to occur.

- *Visibility.* The most important visibility aspects for cul-de-sacs are at a distance from the device itself. Landscaping or other clearly visible provisions should identify the fact that the street is not a through street. Devices should be highlighted with reflectorized paddles or button reflectors.

- *Signing.* Standard signs (such as "Not a Through Street," "Dead End" [W14-1], or "No Outlet" [W14-2]) should be clearly visible at the block entrance to prevent inadvertent entries. Designs with emergency vehicle passage should include standard "Do Not Enter" (R 5-1) signs.

- *Violation Prevention.* In cases where a barrier is placed in the face of even modest community or driver opposition, the lawns, sidewalks, and driveways adjacent to the barrier may require protection by wood or metal posts to prevent circumvention of the barrier.

- *Drainage.* As with diagonal diverters, designs of full barriers can maintain existing curbs and gutters, minimizing costs associated with revising drainage flow.

- *Pedestrians, Bicycles, and the Handicapped.* Designs with emergency passage provision should also provide for these nonmotorized travelers. Without emergency passage, special ramps for bicycles and wheelchairs may be needed if sufficient numbers of them are present. As with diagonal diverters, pedestrian continuity can be aided by extending sidewalks across the end of the barrier.

Midblock Cul-de-Sacs

A cul-de-sac placed within a block, rather than at one end, performs the same function as an intersection cul-de-sac (Fig. 5.9).

A midblock barrier can be especially useful in locations where a high traffic generator borders a residential area. The midblock barrier can permit access to the generator from a major street while protecting the neighborhood from through traffic. Traffic effects, design features, typical construction materials and costs, and legal status are similar to those listed in the previous section. However, it should be noted that there is a greater prob-

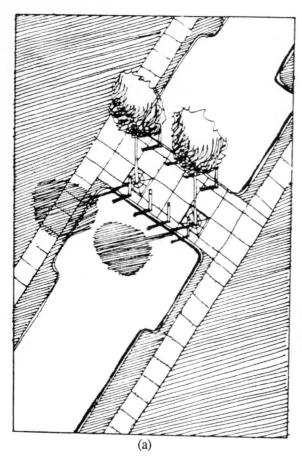

(a)

Figure 5.9   Midblock cul-de-sac.

(b)

ability of motorist confusion when a midblock cul-de-sac cannot be seen from the intersection of boundary streets. Signing as recommended for cul-de-sacs above is advisable
to reduce such confusion.

Pavement Undulations

Pavement undulations and speed bumps are among several physical devices (others being
rumble strips, traffic circles, and raised intersections) which have been used for the primary
purpose of reducing speed. They are raised areas in the pavement surface extending
transversely across the traveled way. Normally they have a height of 3 to 4 inches (7.5 to
10 cm). Length in the direction of travel distinguishes "bumps" from "undulations." Bumps
are abrupt, normally less than 3 feet (1 m) in length, while undulations are more gradual
with lengths of 8 to 12 feet (2.5 to 3.5 m). Bumps have occasionally been installed on park
roadways and private drives and rarely on public streets. A recent innovation, undulations
have been installed on public streets in an increasing number of jurisdictions in the United
States and more widely in Europe and elsewhere (Fig. 5.10).

   There are several reasons why the use of bumps on public streets has been limited.
The major one is a real or perceived question as to their safety. Also, they are not included
in the MUTCD and parallel state control and design guides; as a result, many traffic engineers consider them illegal.

   On the other hand, undulations developed by the Transportation Road Research
Laboratory (TRRL) in Great Britain appear to have promise for effectively reducing speed
and providing better safety characteristics than bumps.

Figure 5.10 Pavement undulation.

By 1983 there were between 150 and 160 undulation units deployed on public streets in California alone and extensive use of the device is increasingly reported in Europe and North America. A survey performed jointly by the Federal Highway Administration and the City of Thousand Oaks, CA, found that 25 responding jurisdictions had installed undulations, 6 more were planning to install them, and 63 others were evaluating the possibility. By early 1986, Thousand Oaks and Pasadena, CA, each had about 60 undulation installations and had *not* experienced any adverse traffic safety incidents related to them.

Melbourne, Australia, has developed *raised paving plateaus* as an alternative to undulations. These are installed at midblock locations on more important local routes. They are constructed from concrete paving blocks about 100 millimeters (4 in) high; each plateau's length in the direction of travel is about the same as the curb-to-curb width of the street; ramps at a grade of from 7 to 8 percent lead to and from the plateau. Such plateaus are reported to be acceptable both to drivers and residents.

The basic purpose of pavement undulations is obviously to reduce speed. However, the actual design of the undulations is critical to their ability to achieve this. Tests have also shown that some designs produce less discomfort at higher speeds than at low ones, in direct contradiction to their purpose. The most successful undulation units have the design features recommended below.

a. *Effect on Traffic Volume.* Undulations usually cause at least small traffic volume reductions on the streets where they are employed. This is natural since the undulations introduce slower speeds and a discomfort factor to the street in question. The extent to which diversion occurs is largely dependent on the configuration of and flow conditions on the area street system rather than on the properties of the undulation installation. However, a series of closely spaced undulations are likely to produce more diversion than a broadly spaced sequence.

b. *Effect on Traffic Speed.* Undulations have been shown to reduce the 85th percentile speed on the average between 14 and 20 mph at the device itself and to also produce substantial reductions in speeds on the road segments between undulations. The extent of speed reduction achieved between undulations is related to the spacing distance between undulations. At spacings under 800 feet (250 m), undulations exert a rather continuous effect on drivers' choices of speeds, but at greater separation distances they have an effect only in their immediate vicinity (much like a stop sign).

c. *Effect on Noise, Air Quality, and Energy Consumption.* When used on low-volume local streets, undulations normally produce small reductions (1 to 2 decibels) in average sound levels both at and between the devices. On busier streets or streets with significant truck volumes, noise levels can increase.

d. *Effect on Traffic Safety.* A 1983 study of pavement undulations by a subcommittee of the California Traffic Control Devices Committee found that between 150 and 200 million vehicle crossings of the 150 to 160 undulations on public streets in the state had taken place without incident. No cases of motorists losing control of a vehicle were reported, and, while a few claims for damages to vehicles allegedly caused by the undulations had been filed, in only one instance had a plaintiff been provided compensation (less than $20). Emergency vehicles, buses, and large trucks must pass over the undulations at relatively slow speeds (under 20 mph) or else significant jolts to the vehicle, discomfort to occupants, and jostling of cargo will be experienced.

Desirable design and location features

- In profile the undulation should have a generally circular arc cross section on a 12-foot (3.5-m) chord with a maximum midpoint height of 3 inches (7.5 cm) and an allowable construction tolerance of plus or minus 0.5 inch (1.2 cm). (This recommended height is less than the 4-inch (10 cm) value recommended in early research reports.) The undulation should extend across the roadway with the last 1 to 3 feet (0.3 to 1.0 m) tapered so that it becomes flush with the gutter pan to maintain drainage flows.
- Undulations should be placed singly. Closely spaced pairs, though utilized successfully by some jurisdictions, do not appear any more or less effective than single undulations.
- Undulations should be spaced approximately 550 feet (165 m) or less apart.
- Undulations should be placed at least 200 feet (60 m) away from intersections and sharp horizontal curves and be otherwise located so they are clearly visible for at least 200 feet (60 m).
- Specific positioning of undulations should consider access to utilities, driveway locations, and existing illumination.
- The undulations should be marked with warning signs at the device and pavement messages in advance. Advance warning signs, advisory speed plates, double yellow centerline marking in the vicinity of the undulation, and pavement markings on the device are optional.
- Unfortunately, major and collector streets which are residential in character are those on which traffic speed is a significant issue. However, undulations should *not* be utilized on these classes of streets because the level of restraint they impose is inconsistent with the functional purpose of the streets.
- Undulations should *not* be used on grades greater than 5 percent.
- Undulations should *not* be placed on primary emergency vehicle access/egress routes nor on important transit routes.

Raised Intersections

A raised intersection is analogous to a midblock pavement undulation as a speed reduction technique. The entire intersection is raised a few inches above the normal grade level, with ramps to conform to the grades of the adjacent streets. This design has been employed extensively in Europe (Fig. 5.11). Some raised intersections have been installed in the United States (notably in Hartford and Seattle), generally as pedestrian safety or convenience measures on shopping streets rather than as a neighborhood traffic control.

In Melbourne, Australia, more than 100 raised intersections were reported in use by late 1986. These "plateaus" are about 100 mm high and cover the entire intersection area with approach ramps sloped at grades of from 7 to 8 percent. (These vertical dimensions are similar to those used in midblock plateaus mentioned in the preceding section.)

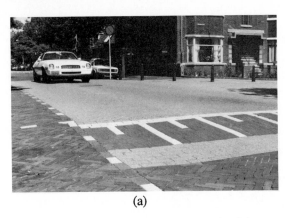

(a)

(b)

Figure 5.11 Raised intersection in residential area.

Rumble Strips

Rumble strips, patterned sections of rough pavement, were developed in the 1960s as a means for alerting drivers to the presence of a dangerous condition or a specific control device. While used primarily on freeways and major streets, they may have been used as a speed reduction device in neighborhoods.

a. *Effect on Traffic Volume.* None.

b. *Effect on Traffic Speed.* The studies which have evaluated the effects of rumble strips on speed show their effect to be mainly at the upper end of the range of acceptable speeds in residential areas.

c. *Effect on Noise, Air Quality, and Energy Consumption.* As is suggested by their very name, rumble strips generate noise when traversed by motor vehicles. Studies of rumble strips have documented only in-vehicle measurements; no measurements for interiors and exteriors of adjacent residences have been reported. However, in recognition that noise generated by the frequent passage of vehicles may be a nuisance, it is suggested that use of the strips for neighborhood traffic control purposes be limited to streets with less than 2,500 vehicles per day.

No significant air quality or energy consumption effects have been demonstrated.

d. *Effect on Traffic Safety.* Studies conducted on major streets show that the strips have had a noticeable positive effect in reducing accidents when placed in advance of a stop sign. Effects in lower-speed residential areas have not been determined. No specific instances of bicycle accidents resulting from rumble strips have been reported. However, cyclists often complain that the types of raised pavement markers sometimes used in rumble strips make riding difficult and unpleasant.

e. *Uniform Standards.* While not treated in the MUTCD, rumble strips are a recognized device in basic traffic engineering reference texts. However, no specific warrants or design standards are given.

f. *Community Reaction.* Negative citizen reactions to the noise levels generated by the strips can be anticipated.

Traversable Barriers

Traversable barriers are special provisions to permit passage of selected vehicles through full diverters or cul-de-sac treatments. They include vehicle passageways across mountable curbs, gaps in diverters or other barriers—often protected by flexible or breakaway stanchions, raised traffic bars, or "undercarriage preventer devices" (Fig. 5.12)—and pas-

Figure 5.12 Traversable barrier.

(a)

(b)

sageways traversable only by vehicles with wide wheel spacings (such as buses, trucks, and fire equipment) or guarded by automatic or manually operated gate devices.

Each of these measures has inherent problems. Any emergency vehicle gap must be kept free of obstruction. Where enforcement is lax, parked cars blocking such gaps are not infrequent. Barriers with open gaps (restriction of passage to emergency vehicles by signs and markings only) are subject to violation by unauthorized vehicles.

As an alternative to simply providing an open gap, some communities have placed mountable curbing across the emergency vehicle passage to discourage motorists. However, any curb which poses a somewhat formidable barrier to normal traffic also slows emergency vehicles. If such vehicles attempt to take a raised curb at speed, the shock of crossing can lead to problems with wheel alignment, dislodge equipment, and pose safety problems for the crew or vehicle occupants. Where an emergency vehicle passageway is provided through a raised barrier device, smooth ramps are preferable to mountable curbs.

Protection of emergency vehicle gaps by means of flexible or breakaway materials is inadvisable. Flexible barriers are equally permeable by violators in private vehicles. Breakaway barriers are less subject to violation; however, debris from broken material poses a hazard to emergency vehicle occupants, including firefighters riding on the exterior of their apparatus. Some communities have placed raised traffic bars in the emergency vehicle openings. However, these have not posed much of a deterrent to drivers determined to violate the barrier. And, when fire apparatus or ambulances traverse them at speed, they tend to dislodge equipment and hazardously jolt firefighters, or threaten the safety of ambulance occupants.

A few cities have employed an undercarriage preventer device in the emergency vehicle passage. This is a wooden or concrete block, usually about 3 feet (1 m) wide, 6 inches (0.2 m) thick, and raised about 6 inches (0.2 m) above the surface of the passage pavement. In theory, emergency vehicles are higher slung than most other vehicles in normal public use. By measuring the underbody clearance of all types of emergency vehicles in use, the projection height for the undercarriage preventer device which will allow emergency vehicle passage but discourage or prevent other vehicles from traversing the barrier gaps can be selected. In practice there are problems. A projection height which can be cleared adequately on a flat roadway surface can cause the same vehicles to bottom-out when the device is placed on a crowned contour. Because crowns vary substantially, projection height must be determined on an individual site basis. Some emergency vehicles differ little in underbody clearance characteristics from the vast majority of automobiles in normal use. Conversely, virtually any private truck and some common high-slung automobiles can clear almost any undercarriage preventer that fire apparatus can clear. Therefore, the undercarriage preventors are not wholly effective.

Calgary has installed numerous "bus gates" that permit large vehicles to cross barriers intended for regular traffic at the end of cul-de-sacs. These bus gates consist of a pit spanned by a grating with bars spaced far enough apart longitudinally to prevent crossing, and bridged by two runways spaced laterally at the track width of buses and other heavy vehicles, but too far apart to be traversed by standard automobiles (Fig. 5.13). Besides

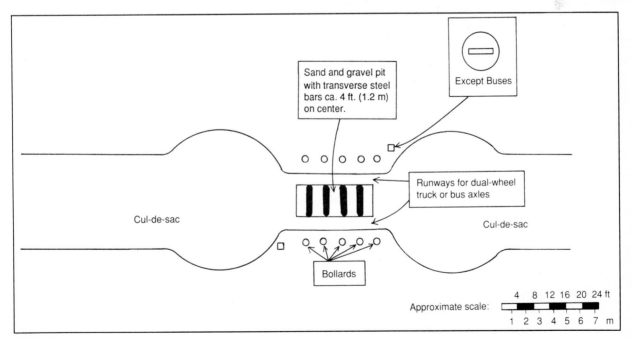

Figure 5.13 "Bus gate," Calgary, Alberta.

buses, fire apparatus, refuse collection vehicles, and other heavy trucks are able to pass through this device.

Gates of the type used in parking facilities, but equipped to open in response to a radio signal sent by an emergency vehicle, have proven problematic because of susceptibility to vandalism. More substantial gates and opening mechanisms — such as an adaptation of railroad grade crossing protection gear — may be vandal-resistant, but involve considerably more installation cost than the simpler parking facility gates.

Manual devices in which the emergency vehicle operator or a crew member dismounts to unlock and open a gate or remove a retractable bollard have been in use in vehicle-free zones in many areas for a number of years. While they work acceptably in simple situations of providing emergency access to individual blocks, they are less satisfactory where an emergency vehicle is attempting to traverse several blocks.

Other Geometric Features Used in Residential Traffic Controls

Among miscellaneous geometric features used on local streets are the following:

a. *Curvature.* Introducing curvatures on a previously straight alignment has been discussed as a physical speed control device. In San Francisco's ill-fated Richmond district traffic control project, chokers were installed on opposite sides of alternate blocks to create a serpentine alignment. However, due to public controversy the devices were removed before performance measures could be taken. Swedish reference sources suggest the possibility of introducing sweeping curvatures or tighter kinks ("knixars") into the roadway alignment as speed control measures. But they warn of possible associated safety problems. Even more drastic is the British "Z-track" concept. Still in the paper stage, this concept involves the use of curbs or other barriers to contort the roadway alignment, actually in the approximate shape of the letter N rather than Z, so that a vehicle must actually back down the crossbar in order to continue.

Australian experience with introduction of curvature into straight streets falls into two categories:

- Reconstruction of the street to align the curb, deflecting the centerline and creating space for landscaping. This has generally proven to produce popular improvements to the street appearance, but is very expensive and seems at best to have a very subtle influence on driver behavior.
- "Slow Points." These are curbed islands or curb extensions protuding into the roadway, leaving a single-lane (sometimes a narrow two-lane) gap, often at an angle to the centerline. A combination of such devices is sometimes used to create an S-shaped path. Done correctly, these seem to have a useful place in the range of geometric modification strategies.

b. *Other Pavement Irregularities.* Valley gutters and rough pavements are two existing devices which tend to control traffic as an unintended byproduct of their presence. Valley gutters may run parallel or perpendicular to the direction of travel. In either case, they appear to be somewhat effective in reducing speeds in their immediate vicinity. Likewise, roads with rough surface, possibly in need of repaving, have a speed reducing effect. In neither case can it be suggested that streets should be designed to include valley gutters and rough pavement in order to reduce speed; however, the effect may be an argument for delaying repaving of purely residential streets.

COMPLETE STREET CLOSURES

Complete street closures are often impossible, because access to the driveways of abutting properties may have to be maintained. Three exceptions to this situation are described here: streets that can be closed for certain hours, especially for use as play streets; streets whose ownership can be transferred to private neighborhood organizations who can then exclude general traffic; streets where all abutting properties also have access via a side street or alley.

Play Streets

Temporary play streets are common in some East Coast American cities, notably Philadelphia and New York, which each have over 150. These streets are temporarily closed during specified hours in the summer. Some communities may use signs only for the temporary closure; others use portable barricades or a combination of both.

The size of these programs speaks both to the level of need and to the capacity of the cities' street systems to function with so many closures. Most of these streets are in low-income neighborhoods, and the main objectives are to reduce accidents and provide youngsters with play space during the school vacation. Many are operated with supervisors and temporary equipment. The surface of the street is marked to facilitate the conduct of various games. They are usually sponsored by block associations or community organizations. Many are on one-way streets. As to traffic considerations, there must be assurance that the street closing will not adversely affect deliveries or parking, or cause problems of diverted traffic. If parked cars are present before the block is temporarily closed, motorists who choose to exit the area during play hours can cause or have problems.

Total closure of streets is another physical control which has been practiced to a greater extent in Europe than in the United States. Generally, the total elimination of automobiles from a street has been limited to central business districts, rather than residential areas. In the United States, such a proposal in residential areas would of necessity be limited to areas where vehicular access to homes and garages is provided for by alternate means. Areas with alleys to provide access might be suitable for such treatment.

Private Streets

Restriction of traffic to residents, their guests, and (in some cases) deliveries, can be accomplished by designating streets as private property. For new subdivisions, this is accomplished by not dedicating the streets for public use, with maintenance responsibility assumed by residents of the area. To convert an existing public street into a private one, the government must legally vacate it with an agreement among property owners to maintain it at their expense. This may also require a special provision for postal deliveries, garbage pickup, and other municipal services as well as utility easements.

Gates have little application on public streets, but are often associated with private streets. Mechanical gates have been used to control access to private parking facilities, and elaborate architectural gates have been used at the entrances to exclusive residential areas. In Windsor Park, Great Britain there has been some success with movable gravity-closed gates which are opened by the car softly striking the gates themselves. The design is rather crude. The driver must judge how hard to hit the gate so as to open it yet not have it bounce back to hit his car. The potential for children being hit by the swinging gates is high. Despite this, the gates have operated to the satisfaction of residents and have been without accidents for 30 years. There would appear to be potential for improving upon this design, assuming that legal concerns can be satisfied.

Closures Where Alternate Access to Property Is Available

In some existing neighborhoods it may occur that all land parcels along a street also have access via alternate streets or alleys. It may then be possible to close selected blocks of such streets.

As an example, the West End district of Vancouver, British Columbia is illustrated (Fig. 5.14). East-west streets are not only fairly closely spaced, but are also paralleled by intervening alleys. As a result, there are never more than two land parcels along a north-south street between a cross street and an alley, and these do not require vehicular access to the north-south streets.

As shown in Fig. 5.14, a number of blocks on north-south streets have therefore been closed to vehicular traffic and converted through landscaping to exclusive use by pedestrians, cyclists, strollers, and—in one case—an outdoor cafe. Front doors of abutting properties remain as before, but residents must park their vehicles on a cross street or alley, or—where available—in off-street spaces. Provision for emergency vehicle entry into these blocks is provided by use of traversable barriers in some, but not all, cases. Fig. 5.15 shows the extensive amount of landscaping, lighting installation, and other amenities which have been accomplished.

In considering such projects, the quality of the alleys involved must be taken into consideration. Inevitably, some traffic will be diverted to alleys for at least one block from the point at which the motorist encounters a closed street. If alleys are narrow, cluttered with intruding structures, or used extensively for parking, they may be inadequate for detouring traffic.

COMMUNITY SERVICE IMPACTS AND COUNTERMEASURES

Regulatory, warning, and guide traffic control devices used in neighborhood traffic restraint have little impact on community services. On the other hand, geometric features used for neighborhood traffic control can affect community service vehicles by blocking their paths or hindering their mobility. Devices of concern include diverters, half-closures, cul-de-sacs, circles, forced turn channelization, and median barriers. Primary concerns are for fire, police, and ambulance services, and for private vehicles traveling in emergency situations. In addition, routine services such as public transit, school transportation services, delivery vehicles, refuse collection, and street and utilities maintenance operations can be affected.

This section examines in detail the effects of traffic management devices on all of these operations, and proposes engineering and design measures which can counter these effects. The objective in residential street traffic control is to create conditions which are characteristic of well-designed modern residential suburban street systems. The services of concern all generally function satisfactorily in such street systems. Therefore, the cautions put forward here should not be viewed as insurmountable obstacles to implementation of neighborhood traffic control schemes. The intention is to alert the analyst to the potential site-specific or systemwide effects, and to ways they can be avoided or made tolerable. Careful planning and engineering can avoid or mitigate the likelihood of serious adverse effects and result in conditions similar to those in well-designed suburban communities.

Fire

a. *Potential Impacts.* Concerns about the impact of control devices on firefighting operations center on two elements: response travel time and maneuvering of equipment during firefighting operations.

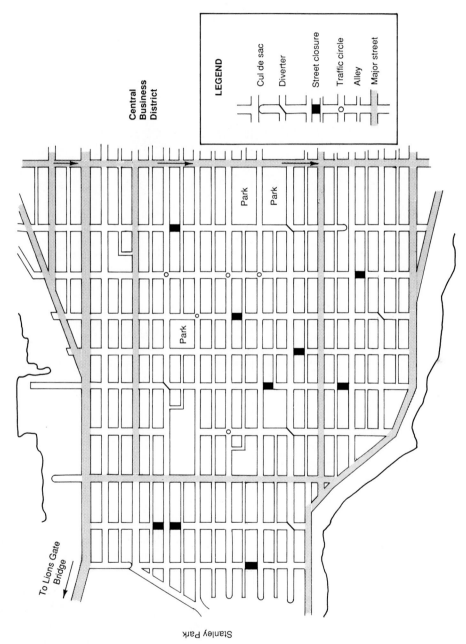

Central Business District

Park

Park

Park

To Lions Gate Bridge

Stanley Park

LEGEND

Cul de sac

Diverter

Street closure

Traffic circle

Alley

Major street

Figure 5.14 West End neighborhood, Vancouver, Canada.

113

Figure 5.15 Street closures in the West End, Vancouver, Canada.

The response time—the time between the start of a fire and the beginning of extinguishment—has a significant relationship to accomplishing lifesaving, minimizing property damage, and the difficulty of extinguishment. Geometric features potentially affect response times in several ways:

- Force apparatus to longer, less direct routings.
- Confine apparatus to busier streets, exposing it to more congestion/delay.
- Lead to apparatus ending up on the wrong side of a barrier from a fire.
- Preclude the good practice of routing multiple firefighting companies via parallel routes.
- Lead to an entire area being temporarily inaccessible to fire apparatus, such as when barrier devices interrupt some streets, and episodic incidents (sewer and water hookups, street repair, tree pruning, and the like) block others, with traffic from such blocked streets jamming the remaining streets.
- Slow down heavy fire apparatus maneuvering through or around geometric features.

On the other hand, if geometric features are designed to facilitate emergency vehicle passage, response time may be improved where such vehicles can travel on little-used streets with no conflict from other traffic.

Firefighting operations also may be adversely affected by geometric features. At the scene of the fire, barrier devices may:

- Interfere with maneuvering and effective deployment of apparatus and equipment, particularly with effective deployment of tillered aerial ladder apparatus.
- Interfere with access to water supply points.
- Complicate diversion of traffic away from the fire scene.

b. *Countermeasures.* Potentially adverse effects of traffic barriers on response travel time can be offset by other improvements affecting travel time or by improving fire detection and alarm transmission. Some possibilities include:

- *Signal preemption.* Hardware permitting emergency vehicles to preempt traffic signals could cut response time along major and collector streets, offsetting increases caused by barriers on other streets.
- *Improved detection.* Reliable, low-cost combustion detection units are capable of significantly reducing detection time. A community considering a traffic management plan involving barriers could require installation of such detection units in all structures to offset the increases in response travel time.
- *Improved alarm transmission.* Upgrade telephone and modern electronic signaling, retransmitting, and dispatching devices, again to offset increases in response travel time.

While the above possibilities hold some promise, the best approach might be to design a neighborhood traffic management system which would minimize its adverse impact on response and firefighting operations. Primary solutions include making barriers traversable, planning the neighborhood traffic management barrier system to minimize interdiction of primary fire routes and operations in the vicinity of large buildings (potential multi-alarm fire sites), and providing additional fire hydrants where barriers compromise accessibility to existing fire plugs.

The potential for traffic management plan interference with fire apparatus accessibility and firefighting operations can be minimized through good planning practices, such as:

- Avoiding the placement of barrier-type devices on the main egress routes from fire stations.
- Using half-closures or one-way street mazes in preference to full barrier treatments where strong traffic control devices are necessary on primary fire station egress routes.
- Minimizing the use of full barrier treatments in close proximity to potential multi-alarm fire sites — multistory apartment buildings, places of public assembly, and so forth.
- Installing new fire plugs on the "dry side" of any barrier device which does not have an emergency vehicle opening.
- Providing up-to-date maps showing neighborhood traffic control features not only to stations within the community but also to those in nearby jurisdictions who may be called to offer assistance.
- Encouraging each fire station to plan response routes to all parts of the area, taking barrier devices into consideration.

Police

a. *Potential Impacts.* Concern about the effects of barrier-type devices on police functions centers on four topics:

- Barriers make it more difficult for police to patrol an area, thereby decreasing police surveillance.
- Barriers tend to hamper patrol car pursuit of motorcycles, motorscooters, bicycles, and suspects fleeing on foot.
- Use of large numbers of barriers in the city as a whole or in one or several adjoining neighborhoods could adversely affect police response time to emergency calls.
- The ability to use streets paralleling major or collector streets as an alternate route in cases of blockage of the main street due to fire, construction activity, or special events is a police concern.

Available information, however, lends minimal support to these concerns. Relative to the patrolling issue, studies have demonstrated that blocks with lower accessibility (characteristic of situations where diverters and cul-de-sacs are employed) tend to experience less residential crime than blocks with higher accessibility, even though exposed to similar crime-related variables. Comparison of residential crime statistics before and after implementation of traffic management plans lends no support to the hypothesis that neighborhood traffic management would lead to greater crime rates due to inhibited police patrolling.

Experience with neighborhood traffic management barriers gives slight support to the notion that the barriers might pose obstacles in "hot pursuit" situations, but there is little evidence of police identifying traffic barriers as the cause of failure to capture a suspect. On the other hand, there are reports of situations where the capture of suspects fleeing by car was aided by barrier systems because drivers unfamiliar with the area became lost in the barrier maze. Moreover, many communities discourage the "hot pursuit" concept, preferring multi-unit interdiction. "Before" and "after" data compiled by the police department in a city which installed an extensive neighborhood traffic scheme indicate that the presence of barriers and other traffic control devices did not have any significant impact on overall police response time.

b. *Countermeasures.* Despite the lack of evidence that police functions are significantly impaired by barrier-type devices, communities may wish to take steps to reduce potential adverse impacts to a minimum. The strategies for doing so are similar to some of those listed for fire services: improving alternate routes, minimizing the use of full barrier treatments near stations, utilizing traversable barrier designs, and providing up-to-date maps showing neighborhood traffic control features.

Ambulance Services

Both potential impacts and strategies in dealing with ambulance services in neighborhood traffic management plans are similar to those for fire and police. The placement of barriers on the immediate egress routes of ambulance operating bases and on the immediate access routes to hospital emergency rooms and emergency clinics should either be avoided, or the barriers should have traversable passageways suitable for ambulances. Avoiding primary ambulance routes is a key consideration in the use of undulations and other pavement roughness treatments, because ambulance patients may be extremely sensitive to vibrations during the trip.

Deliveries and Refuse Collection

Regularly routed vehicles for deliveries (postal service, milk, newspapers, and so forth) and refuse collection can have their routes adjusted to operate in an efficient and continuous pattern within the constraints imposed by a system of barriers. Only cul-de-sacs significantly decrease efficiency by forcing vehicles to backtrack over previously covered ground. For occasional service and delivery trips (such as furniture vans, appliance installations, phone repairs, meter readings) the barrier scheme poses a problem to drivers unfamiliar with the area, just as it does to visitors. Signs giving adequate advance warning can mitigate this problem. Also, up-to-date maps can be distributed to the affected businesses to help them instruct their employees on available routes.

Transit

Since public transit service operates primarily on major and collector streets unobstructed by barriers, minimal interference with these operations generally occurs. Transit vehicles may be exempted from some restrictive controls by sign notice if it appears safe to provide such an exemption. Barrier systems may pose more of a problem to paratransit vehicles, dial-a-ride operations, and school bus service, which usually travel to some extent on the residential street system. Careful design of the barrier system with respect to school bus operations tends to minimize interference, although some relocation of pickup points may be necessary. Dial-a-ride and other paratransit users can adjust their operations in much the same manner as regular deliveries and are likely to be minimally impacted by barriers.

Maintenance

The potential impacts of traffic management on maintenance operations schemes must be considered in the planning stage so that appropriate adjustments can be designed, and cost impacts of operational changes can be assessed. However, interference of barrier devices with normal maintenance activities is typically minor. Northern cities have reported that diverters, cul-de-sacs, and speed bumps complicate removal operations in heavy snow conditions. Where sewage system manholes are typically located in the center of intersections, diagonal diverter barriers must allow for normal operation of sewer flushing equipment. Diverse examples of similar kinds of problems have been reported; none are of a particularly serious nature.

VIOLATIONS OF TRAFFIC BARRIER DEVICES

Barrier devices, particularly when designed so that their intent is obvious (such as without landscaping to conceal the formerly open street), may frustrate motorists and create a considerable level of driver resentment. Some drivers respond with behavior that reflects this resentment. A few resort to vandalism, but more prevalent behavior is violation of the device by driving through or around it. The actual extent to which this occurs depends on how easy it is physically to violate the barrier, the amount of advantage the driver gains by this compared to exercising other options, and general expectations regarding enforcement and the consequences of being caught in violation.

It is difficult to generalize about the rate of likely violations for various forms of devices. Barriers with open paved gaps, such as semidiverters or diagonal diverters, and cul-de-sacs with unprotected emergency vehicle passages are obvious targets for violation.

In Berkeley, CA, where some 70 diverters and cul-de-sacs have been deployed as part of the citywide neighborhood traffic management plan, barriers with open emergency vehicle passages experienced violation levels on the order of 5 to 7 percent of the traffic volume formerly using the street. Counts of violations of emergency vehicle passages protected by undercarriage preventers showed no significant differences in the violation rates. However, the undercarriage devices were employed only at locations where gaps were initially observed to have high rates of violation. Even with such levels of violation, it is clear that the barriers are highly effective in reducing through traffic volumes on the streets on which they are deployed.

Devices or combinations of devices which tend to entrap motorists inside neighborhoods are ones most likely to be violated, particularly by persons encountering them for the first time. For this reason there may be a tendency toward higher violation rates for cul-de-sacs and midblock closures unless they are designed to be violation proof. For similar reasons, half-closures which prohibit exits from a block rather than entries to it are to be avoided. High violation rates can also be expected at sites where the alternative route involves significant out-of-direction travel or passage through heavily congested streets and intersections. Violations of barriers on interior neighborhood streets is more likely than of those on the periphery.

The best way to avoid violation problems is to landscape the devices so well that the roadway on the other side is hardly visible and looks as if there never had been any connection between the two street segments separated by the barrier device. There is some evidence that even a moderate landscape treatment reduces the tendency for motorists to violate the control. Bollards may sometimes be necessary to prevent residents' lawns from being used as a detour route. A modicum of enforcement is also required; otherwise violations may become pervasive.

6
Implementing Neighborhood Traffic Controls

INTRODUCTION

The way in which neighborhood traffic control strategies are implemented can be as important to their eventual success or failure as the substance of the strategies themselves. Implementation should be considered not as a step but as a *process* requiring careful planning and documentation, public notice, evaluation, and possibly refinement of the strategies. Such a process calls for the same attention to detail and the same thorough consideration as the initial planning effort.

This chapter discusses the major topics which need to be considered in designing and carrying out an implementation process for neighborhood traffic controls. In the next section, legal issues are addressed, with emphasis given to the kinds of findings and documentation local governments may wish to prepare in order to assure that their neighborhood traffic control plans are on solid footing. The third section of the chapter addresses issues in the design of an implementation process, including public notice and some of the key choices that usually must be made about how to proceed. The fourth section considers evaluation of neighborhood traffic control plans. Finally, the maintenance of the traffic control plan is discussed.

LEGAL ISSUES

The implementation of neighborhood traffic control schemes, particularly large-scale ones or those that involve street closure or substantial traffic diversion, may raise issues about the responsible jurisdiction's authority to take such actions. As discussed in Chapter 3, a local jurisdiction can do much to assure that its traffic control plans are legally sound by carrying out a thorough planning process and documenting its findings carefully. These steps can help substantiate that the jurisdiction is acting reasonably and in accordance with applicable law. In actually implementing the traffic control plan, however, several additional legal considerations may be involved. For example, the measures or devices used to effectuate traffic control may be subject to state requirements as to design and/or application. Legal questions may also arise due to restrictions of access caused by the plan, its environmental impact, or concerns for tort liability.

The sections that follow describe some of the legal issues that a local jurisdiction may need to consider in implementing neighborhood traffic controls. It is beyond the scope of this book to report on the numerous legal requirements of the various U. S. states or of foreign jurisdictions, however, and readers are strongly advised to work closely with their attorneys in developing and implementing neighborhood traffic control strategies.

Establishing Implementation Authority and Demonstrating Reasonableness in Its Exercise

In the United States, the right to regulate and control the operation of motor vehicles on streets and highways is a police power of the state. As discussed in Chapter 3, states routinely delegate a certain degree of this authority to local jurisdictions for control of local streets. The conveyance of such authority varies from state to state and, in some states, from jurisdiction to jurisdiction (for example, charter cities and general law cities may have different powers). Delegation may be limited to specific permitted actions, such as to close streets, utilize state-authorized traffic control devices, and/or restrict certain kinds of traffic (such as trucks) as to time or place. In other instances, an omnibus clause may empower jurisdictions, in addition to powers specifically delegated, to maintain and enforce such reasonable ordinances, rules, and regulations with respect to traffic as local conditions may require, so long as these actions do not conflict with powers reserved to the state. In some jurisdictions, additional authority to regulate traffic may derive from state authorizations or mandates to implement the circulation element of a duly adopted General Plan, or more broadly from authorization to take such steps as are deemed necessary to protect the public health and welfare.

Particularly in those areas where local authority to regulate traffic is narrowly defined, jurisdictions should take care to select implementation strategies which are consistent with the permitted actions. For example, a jurisdiction with authority to utilize state-authorized traffic control devices probably should develop its neighborhood traffic control plan utilizing traffic control devices recognized in the Manual on Uniform Traffic Control Devices (MUTCD), published by the U.S. Department of Transportation, or in parallel state manuals. An area with authority to close streets may be able to utilize geometric features closing at least portions of streets, but should take care to follow all procedural requirements upon which street closure is conditional: typically, public notice and hearings are necessary, and in some jurisdictions the abutting property owners must agree to the closure. Selecting devices and following procedures that are consistent with the traffic control powers delegated to the local jurisdiction are an important start in avoiding challenges which allege lack of authority or improper use of authority.

While local jurisdictions usually can demonstrate some basis of legal authority to implement traffic management actions, they may be challenged on the grounds that the actions are arbitrary, capricious, and unreasonable and therefore not a legitimate exercise of the police power. One strategy that is useful in demonstrating that the police power was reasonably exercised in neighborhood traffic management cases is to set forth in writing the objectives of the plan, the facts and other evidence that support the actions being taken, and the procedures used in deciding to proceed. This written record can help citizens to understand the reasons for the traffic control plan, as well as provide an element of protection in case the action is challenged in court.

The following kinds of findings could be made as part of an ordinance or resolution adopting the traffic control plan, or could be provided in supporting documentation:

• The need for action should be documented. Findings based on both hard evidence, such as traffic accidents and traffic counts, noise and air pollution measurements or calculations, and on citizen concerns over litter, traffic speeds and volumes, child safety, and general traffic nuisance, can be made by the government body authorizing the traffic control actions. Findings from a traffic survey can be particularly useful. Speed profiles and average daily traffic volumes (ADTs), as well as traffic composition (percent of trucks, motorcycles versus autos, "through" versus "neighborhood" traffic) can be compared to the description of the intended use of the residential streets in question; this provides a basis for establishing that the traffic conditions observed are not consistent with the intended function of the street and, hence, that cause for action exists.

• The objectives served by implementation of the traffic control plan should be clearly stated. In most cases, the objectives would include protection of residential areas, reduction of through traffic on residential streets, and so forth. It is useful to tie these objectives to established policies of the jurisdiction. In particular, the relationship of the traffic control plan to the overall transportation plan for the area should be identified. The jurisdiction should cite provisions in the circulation element of its General Plan that support restriction of traffic on residential streets, and should identify the streets which are designated to carry traffic. In some jurisdictions, such citations may help to justify the traffic control actions as an implementation mechanism for the General Plan.

• Alternative traffic control measures attempted or considered can be identified as part of the findings. There is no legal requirement that local jurisdictions who otherwise are operating within the scope of their authority select the least onerous means of addressing a problem, so long as the selected means is rationally related to the problem at hand. Nevertheless, consideration of alternatives, and prior attempts to control traffic through use of stop signs, increased surveillance, and other less severe traffic control measures can help demonstrate that the local jurisdiction is acting reasonably.

• Inconvenience to the traveling public should be accounted for explicitly, either in the authorization for the plan or in supporting documents. Demonstration that the local jurisdiction took into consideration the circuity of alternate routes is an indication of reasonableness of the decision process. Documentation that access for emergency vehicles was taken into consideration when designing the strategy also can be very important. Jurisdictions generally are free to make a finding that effects on the traveling public are insignificant; that they are mitigated by improvements elsewhere (that is, by traffic flow improvements on streets designated for traffic, or by provision of park-and-ride lots, transit service, and so forth); or even that they are outweighed by the importance of residential protection. Note that in some jurisdictions, the latter findings may require that a formal environmental study be prepared.

• Consultation with affected interests also should be documented. The fact that the local jurisdiction widely publicized and held public hearings or employed other forms of community involvement in determining whether to take action and in deciding how the strategy was to be worked out is an important demonstration of reasonableness.

This approach to documenting the reasons for the traffic control actions places heavy emphasis on the importance of a thorough planning program—not just as a vehicle for developing the traffic management scheme itself but as a program which lays solid groundwork for defense against legal challenges that may ensue.

Conformance to Traffic Control and Design Standards

Even if the authority of a jurisdiction to control traffic in neighborhoods is clearly established, questions may arise about the means by which it may do so. In particular, the types of devices or designs that can be used to control traffic may be at issue. There also may be questions about the need to follow "warrants," or established guidelines for the application of certain devices.

In most states in the United States, traffic control devices are required to some extent to comply with the MUTCD or with parallel state manuals, which specify both designs for the devices and warrants for their use. There is considerable variation in the wording and intent of the legislation or regulations calling for such compliance; in some states compliance is mandatory, while in others varying degrees of discretion are permitted.

Compliance with the MUTCD can be problematic because many of the devices commonly used in neighborhood traffic management schemes—diagonal diverters, semi-diverters, retrofit cul-de-sacs, speed control circles, undulations—are not addressed in the MUTCD. Particularly in states where compliance is mandatory, there is the concern that, upon court challenge, removal of the devices may be ordered because they are not found in the MUTCD or the state manuals. There is further concern in such states that motorists involved in accidents may contend that the use of a noncomplying device amounts to a negligent act on the part of the local jurisdiction which was contributory to the accident.

The most straightforward response to these issues would be for the MUTCD and parallel state traffic control manuals to be updated to include specific standards and guidelines for neighborhood traffic control devices—diverters, semidiverters, retrofit cul-de-sacs, undulations, and the like. An alternative direction is to define some of these devices as "geometric features of the road" rather than as "traffic control devices." One state which has taken this alternative action to legitimize such devices is California.

When statute permits the local jurisdiction to exercise discretion in the use of traffic control devices and designs, the MUTCD or other manuals are generally admissible only as evidence of the standard of care. A study of relevant cases found that the MUTCD was considered "as neither an absolute standard nor as scientific truth but as illustration and explanatory material *along with other evidence* in the case bearing on ordinary care."[1] "Other evidence" might include documentation of the need for the particular type of device or design, as well as a careful assessment of its expected performance.

Tort Liability

Many communities hesitate to implement traffic management schemes because they fear lawsuits by drivers, passengers or passers-by who may be injured in traffic accidents involving (or simply near) a neighborhood traffic control. Such liability exposure can be minimized by basing the neighborhood traffic scheme on authorized traffic control devices and street geometric features for which there are recognized standards of practice. For example, a traffic diverter might be created using standard curbing, median designs, directional signs, and roadway markings.

When a plan utilizes features not clearly covered by the MUTCD or other standard sources, the local jurisdiction (and responsible staff members) can take steps to reduce

---

[1] Thomas, Larry W., *Liability of State and Local Governments for Negligence Arising Out of the Installation and Maintenance of Warning Signs, Traffic Lights, and Pavement Markings*. NCHRP Research Results Digest No. 110, April 1979; emphasis added.

potential liability by assuring that the need for the features was clearly established in writing, and that a thorough evaluation of their potential impacts was conducted. Some jurisdictions have developed local design standards and warrants for the installation of such devices as diverters and traffic circles. In any case, a traffic control device or scheme should be designed so that a reasonable driver acting reasonably and exercising ordinary care would be able to readily perceive the intent of the device and safely negotiate that area of the street system.

Other Legal Requirements

In some jurisdictions, additional legal requirements must be met in order for a neighborhood traffic control plan to be valid. In particular, some states in the United States require formal studies of any action that could have a significant effect on the environment. Neighborhood traffic control plans that divert substantial amounts of traffic potentially could have such an effect: if traffic flow on alternate routes worsens, for example, air pollution and energy consumption could increase due to additional stops and starts and lower speeds. Longer routes could produce similar effects. Thus an environmental study may be needed.

Other states or local jurisdictions have requirements for public notice and hearings for actions that change circulation plans; still others require review by the duly designated planning commission. In general, local jurisdictions should take care to comply with all such requirements, since failure to do so could be grounds for a lawsuit.

Challenges Based on the Plan's Impact

Even when there is no question of lack of authority, or of noncompliance with standards or other legal requirements, neighborhood traffic control actions are sometimes challenged by opponents. One ground of action is that property owners' access has been limited. In general, however, unless access to the property in question has been denied completely, the courts have not considered the inconvenience suffered sufficient to challenge the diversion; the courts generally have treated the inconvenience as "an incidental result of a lawful act."[2]

Another challenge to traffic restriction may come from travelers, who may complain that they are discriminated against by a neighborhood protection scheme. Such complaints in the past have been made on the ground that the injured party has been denied equal protection as provided by the 14th Amendment to the U.S. Constitution. However, in a case dealing with resident permit parking, the U.S. Supreme Court said:

> A community may also decide that restrictions on the flow of outside traffic into particular residential areas would enhance the quality of life, thereby reducing noise, traffic hazard, and litter. By definition, discrimination against nonresidents would inhere in such restrictions.[3]

The Court thus cast substantial doubt on the likelihood that equal protection could be successfully used to challenge an action restricting traffic merely on the grounds that nonresidents are treated differently from residents.

In general, challenges to otherwise authorized traffic control schemes on the grounds that they cause incidental inconvenience to some parties are likely to fail; a community may divert traffic and partially restrict access, but still successfully withstand a legal chal-

---

[2] See, e.g., *Mackie v. City of Seattle*, 1978. 576 Pacific Reporter, 2d., 414.

[3] *County Board of Arlington v. Richards*, 1977. 434 United States Reporter 5.

lenge. Tests of sufficient police power and reasonable exercise of such power must still be met, of course.

ISSUES IN THE DESIGN OF AN IMPLEMENTATION PROCESS

In addition to clearly documenting the basis for the neighborhood traffic control plan and assuring that all legal requirements have been met, it is important to develop a systematic procedure for putting the plan into effect. Critical components of such a procedure include public notice, the choice of temporary or permanent installations, development of a financing plan, implementation phasing, and timing of installations. Each of these components is discussed below.

Public Notice and Involvement

Public involvement in the implementation of a neighborhood traffic control plan largely consists of citizens receiving information on how plans will be implemented. The technical staff assumes the duties of informing the citizens of plans and schedules, in order to minimize surprises. Staff also may wish to create a mechanism for citizens to report problems created by work in progress, e.g., by listing a phone number where questions and comments can be relayed to staff.

Publicity about the adopted plan's features and its construction schedule are important components of implementation. If residents and motorists are rudely surprised by abrupt changes in the street system, the immediate result can be erratic and illegal behavior such as dangerous driving maneuvers or outright vandalism. Maps showing features of the plan and its construction schedule should be distributed to residents at their homes, to commuters at their places of employment, and to all organizations operating routed services and deliveries in the city. Notices warning of traffic control changes and dates of construction should be prominently posted around the control sites several days before construction takes place. Where barriers are to be constructed on internal neighborhood streets, similar warning notices should also be posted at the neighborhood entry points and left standing for at least a week after construction is complete.

It is especially important that organizations providing public services, including police, fire, public transit and school bus services, garbage collection, and mail delivery, be fully aware of the implementation schedule. All of these organizations should of course have been participants in the development of the neighborhood traffic control strategies, and alternate routes should have been worked out so that disruption to their functions is minimized.

Since local law enforcement personnel will usually have a major role to play in the success of the new traffic control scheme, the following steps should be considered by staff responsible for implementing the neighborhood traffic control schemes:

- Hold pre-implementation briefings for patrol officers indicating (a) how the plan will affect the performance of their normal duties, (b) why the plan is being implemented, (c) what is expected of them in enforcing traffic laws related to the new controls, and (d) what the expected construction schedule will be.
- Ask police to issue citations for violations to motorists performing radical avoidance maneuvers (such as driving over curbs, sidewalks, and lawns or landscapes) or vandalism and, for less serious violations, to issue warnings initially and citations after a few weeks. The objective is to create an immediate, pervasive impression that traffic laws relating to the devices will be enforced by police and that there is a substantial likelihood that violations will be observed and prosecuted.

- Ask police to enforce regularly no-parking zones in the vicinity of neighborhood traffic control devices. These zones are frequently vital to operational safety and emergency vehicle access.
- Have an informal meeting with local magistrates to inform them of the purpose of the program, the planning process that has been followed, the legal basis for the devices, and the planned enforcement program. Judges are likely to react more favorably when informed in advance of such changes than when their introduction to the program is a series of irate motorists storming the courtroom to protest citations at "strange" traffic controls.

Temporary versus Permanent Devices

The choice of whether to use temporary or permanent devices in initial installations of diverters, semidiverters, cul-de-sacs, circles, and any other devices involving substantial construction involves many trade-offs. The major arguments in favor of temporary devices are that they are usually low cost and easy to modify. The low cost may make it possible to implement traffic controls over a larger area than would be financially feasible if permanent installations were chosen. The ease of modifying temporary devices allows them to be used in situations where the effectiveness or potential side effects of the traffic control scheme are uncertain; the scheme can be considered "experimental," and changes can be made relatively easily, if they prove necessary. Individual installations can be upgraded after they prove successful and as funds become available.

On the negative side, the use of temporary devices may create technical, legal, aesthetic, and political problems. The fact that a temporary device is "disposable" may create a temptation to install it without as thorough an evaluation as would be done for a more costly and permanent device. Furthermore, in order to minimize costs, there may be a tendency to use materials, signs, and markings which may not be in conformance with good design practice. Either could lead to adverse liability situations.

Equally important, if "temporary" devices are installed, issues may never be truly settled. The ready possibility of change may encourage opponents to continue the controversy. Moreover, the temporary devices may offend some residents' aesthetic sensibilities and lead them to oppose the plan because of the devices' unattractiveness. Vandalism and disobedience of temporary devices also has been a problem in some communities.

Permanent installations, by their very nature, seem to command more driver respect, hence better obedience and less vandalism. Residents more readily accept permanent (especially landscaped) devices as enhancements to the beauty of their neighborhoods, whereas temporary materials may be regarded as eyesores. Because permanent installations involve sizable funding commitments, however, professionals and the public need to be sure that they have the "right answer" before proceeding. Especially in complex traffic management schemes, it is very likely that at least some modifications will prove necessary after experience with the new traffic patterns has accumulated. This can make a permanent installation financially risky.

Choice between immediate permanent implementation or initial use of temporary devices should be based on the individual community's situation. In general, temporary installations might be favored in cases where plans are extensive and complex (where the possibility of some planning error is high) and/or where funds are short. Where temporary devices are selected, careful attention to their attractiveness and conformance to traffic control and design standards is a must. Efforts to present an attractive appearance, even with low budget temporary devices, are rewarded.

It also is important to establish a budget and timetable for replacing temporary devices with permanent ones. Without a clear commitment of funds and clear decision points, "temporary" devices may end up being left in place for years.

Financing Implementation

Depending on the extent of the traffic control scheme and the types of devices to be used, the costs of neighborhood traffic control implementation can range from minor to substantial. While both costs and their financing should have been discussed as part of the planning process, usually the details are worked out as implementation draws near and plans are finalized. Both a carefully developed cost estimate and a clear financing plan are needed.

The sources of funds for implementation will vary with local circumstances and the kinds of restrictions that may apply to the use of certain monies. In the United States, general funds most often have been used to pay for neighborhood traffic control schemes, although in some states and under some circumstances fuel taxes, motor vehicle taxes, parking revenues, and other transportation funds can be utilized for this purpose. Occasionally, elements of neighborhood traffic control have been funded by commercial developments whose generated traffic otherwise would have adversely affected the area. In addition, benefit assessment districts or other neighborhood-based funding may be a possibility, especially in affluent areas. Lower income neighborhoods may be able to use community development funds or other grants to help pay for traffic control programs.

While the major costs of implementing a neighborhood traffic control scheme are the devices and their installation, attention also should be given to the extra costs of enforcing the new traffic requirements, especially in the first few months after implementation, and to the costs of monitoring, evaluating, and fine tuning the plan. Explicit budgeting for these cost items is extremely important; unless they are in the responsible staff's budgets and work programs, they may get short shrift.

Incremental versus One-Step Implementation

Devices in individual neighborhoods should be constructed or erected as nearly simultaneously as available resources permit. But if the plan encompasses a large district and involves a significant number of devices, an incremental, neighborhood-by-neighborhood approach might be considered. Indeed, limitations on funding and staff time may necessitate the gradual phasing in of a large plan.

There are pros and cons, beyond resource questions, associated with either approach. The incremental approach allows staff to devote more attention to the details of individual installations and to work closely with neighborhood representatives to make sure that the plan is functioning properly. Also, lessons learned in "early action" neighborhoods can be applied elsewhere, so that repetition of mistakes can be avoided. Yet from a community-wide perspective, an incremental approach may lead to a lengthy period of turmoil as traffic adjusts and readjusts to a continuing series of changes in street conditions. There may be controversy over choosing one neighborhood for early relief of traffic problems and leaving others to later dates. And public reactions to temporary adverse impacts of an early implementation increment can derail a plan at the outset, even though a later staged step would have eliminated the impact.

One-step implementation can avoid issues of favoritism, can reduce the problem of repeated traffic shifts, and can in some instances allow for economies of scale. On the other hand, massive changes in traffic conditions resulting from implementing traffic controls in several neighborhoods at once can create traffic flow and traffic control problems, and may lead to considerable controversy. The sheer size of the program may make it difficult to coordinate construction, monitor performance, provide effective enforcement, and handle citizen inquiries. The visibility of a large scale program can raise the "stakes" unduly; a large program can become a target for political opposition.

The choice between an incremental or a one-step implementation approach is a key judgment call. In most cases, it is likely to be a choice made by local elected officials rather than staff.

Timing of Installations

A practical implementation tactic is to install neighborhood traffic control devices at a time when the least number of drivers and residents are likely to be around. Where possible, staff should take advantage of the "off-season" in a resort or tourist area and of summer vacations in a university town. Year-round residents will have a chance to adjust to the changes during off-peak traffic conditions. Part-time residents and visitors will be confronted with a fait accompli when they arrive.

Not all communities have the advantage of an off-season for traffic. But the converse to the principle applies everywhere—avoid implementation in peak traffic seasons (for example, Christmas shopping season near downtown and shopping centers).

EVALUATION AND FINE TUNING

Thorough evaluations of how neighborhood traffic control measures actually perform have been the exception rather than the rule. In most cases, if the measures implemented satisfy the persons who originally perceived a problem, and if no significant opposition surfaces or serious operational problems result, the program is judged to be successful. Conversely, controversy or accidents may result in a judgment that the measures are a failure. In short, public acceptance and lack of complaints are the most common determinants of the future of neighborhood traffic controls.

Then why evaluate impacts? First, evaluation of technical performance and community perceptions is needed to provide a reasonable basis for decisions to keep or abandon a plan. Actual performance and impacts are often quite different from what is perceived by the neighborhood residents or others affected by the devices. Public reaction is often shaped by first impressions and observations of erratic initial performance characteristics. A formal evaluation can clarify issues, bring the more stabilized long-term performance characteristics into focus, and spotlight hidden gains and losses which may be significant. Favorable claims and adverse allegations regarding traffic safety and congestion impacts must be evaluated objectively lest proponents succeed in perpetuating poorly performing schemes or opponents prevail in having reasonable and effective schemes removed through the political process or legal action.

Second, evaluation makes modification possible. Decisions made without evaluation are typically all-or-nothing—retain the scheme or abandon it. Evaluation can point to opportunities for modifying a traffic control plan to make it perform its intended function better or to lessen adverse impacts. It can also be used to determine if the plan should be expanded both in terms of devices and geographical area.

Finally, evaluation can advance the state of knowledge about neighborhood traffic control, identifying problems which might be avoided in future applications as well as opportunities which might be pursued in ensuing applications. There is much to be learned from experiences with particular devices, approaches, and techniques. Only with thorough, well-documented evaluation of impacts will the state of knowledge in the field of neighborhood traffic control mature.

Chapter 5 presented information on the likely effectiveness and impacts of various traffic control devices and features. This section discusses issues and procedures for evaluation and fine tuning of neighborhood traffic control implementations.

Evaluating the Impacts of Neighborhood Traffic Controls

An evaluation of the impacts of neighborhood traffic controls should start with the question: do the controls fulfill their intended purposes? As noted earlier, the intended effects of residential traffic control may include:

- Reducing traffic volume, particularly through traffic
- Reducing traffic speed
- Excluding undesirable traffic (such as heavy trucks and speeders)
- Reducing accidents
- Improving the appearance of the street
- Creating public space for nontraffic uses such as gardens, play areas, pedestrian amenities, and similar features
- Improving the residents' perception of the street environment as a safe place, especially for children and other vulnerable users

Some of these effects are easily evaluated through a "before-after" traffic study. Basically, data collected during the planning stage constitute "before" conditions which can be compared to parallel measures of conditions "after" implementation to determine changes resulting from the control scheme. Traffic volumes, vehicle mixes, and speeds, for example, can be measured both before and after the implementation of neighborhood traffic controls. Accident data can be taken from local records. Although in most instances the number of accidents in a six month "after" period (or even in a two- or three-year "after" period) will be too low for a statistically significant analysis of before-after changes, the data nevertheless may be indicative of certain kinds of problems (or may suggest benefits).

Other intended purposes of neighborhood traffic control, involving perceptions of improved appearance, safety, and so on, are best evaluated with public input. Several approaches are available: the local staff could keep a log of citizen reactions; staff could interview a sample of residents, local officials, and other interests or conduct a small survey; the planning commission, traffic committee, or other public group could hold a hearing or workshop to solicit public feedback on the matter.

Evaluation should go beyond the question of effectiveness in fulfilling the plan's primary intentions, however. In particular, any negative impacts of the plan as implemented should be identified. For instance, as Chapter 5 noted, certain traffic control devices may inspire defiant reactions on the part of a few drivers; these reactions may be more disturbing to residents than was the original traffic impact from which the device was intended to shield them. Identification and analysis of such effects is important both in developing a balanced assessment of the plan, and in helping to identify modifications which could offset negative effects.

Other undesirable side effects may include:

- Circuitous routes and longer travel paths
- Increased noise, air pollution, and fuel consumption from more stops and starts, slowing for turns, and lower speeds
- Reduced emergency and service vehicle access or increased response time
- Confusion to motorists who are unfamiliar with the new layout or the control devices of the changed street network
- Diversion of traffic to other residential streets where the traffic and its impacts are equally undesired
- Adverse impacts on shops and other businesses which may be located within the area, such as increased turnover or reduced business volumes

Again, some of these effects can be measured directly. For example, in planning the traffic control strategy, technical staff should have predicted the amount of traffic that would be diverted, and the streets to which diversion is likely to occur. Assuming that "before" data on traffic volumes and speeds on these other streets were collected, "after" studies can be conducted on these streets outside the area benefiting from controls to verify the extent of diversion. It is important, as well, to identify any other consequences of the diversion, some of which may not have been fully anticipated.

Technical staff can follow up on such matters as increased emergency response time and increased costs of providing pickup and delivery services by working with police and fire department personnel and other service agencies, who may be willing to keep records on these matters. Of course, early involvement of emergency and service personnel can minimize such adverse effects, by assuring that control devices are traversable by emergency vehicles whenever possible and by allowing service personnel to develop alternate routes.

A log can be used to record motorists' responses to the changed traffic patterns. Technical staff also may wish to use field observations to identify motorist responses in suspected "trouble spots". Also, the workshop or hearing discussed earlier could be useful in allowing a broad range of affected interests to express their reactions and describe their experiences.

As noted in Chapter 3, improvements to the area's transportation services can sometimes be implemented as a means of mitigating adverse effects of neighborhood traffic controls. Such improvements may include: improving traffic flow conditions on nonresidential major and collector streets so that these become the routes of choice for nearly all drivers, building sufficient parking in commercial areas so that traffic attracted to them does not spill over into residential streets in search of parking, and encouraging the use of alternative modes in order to reduce overall traffic. When such mitigation measures have been implemented as complements to the neighborhood traffic control plan, they also should be evaluated. However, some of these measures (especially parking expansion and alternative modes) may not be fully in place or fully effective within the time frame of evaluating neighborhood traffic controls. In such circumstances, the evaluation might simply focus on the amount of progress that has been made in implementing these measures, and might report any preliminary data on effectiveness.

In addition, evaluation includes consideration of other measures not studied in the planning stage. Some of these measures may be relevant solely on an "after" basis (such as incidents in which traffic controls interfered with emergency vehicle operations); others involve "before" and "after" comparisons of information which was not explicitly used in the planning process, but is affected by the plan and perhaps is an objective of the plan (such as changes in residential property values). Analysts should take care that all important measurements of perishable data do get taken, even if some of these are not needed or useful in the initial planning stages.

Experience to date provides some guidance on the kinds of impacts that might be looked for in the evaluation of specific types of neighborhood traffic controls (see Chapter 5). For example, regulatory controls, which involve the use of markings and signs to inform the driver that a specific action is not permitted while not physically preventing the action, are more easily violated than most geometric features. They are most effective in areas where general respect for all types of traffic control is high, where there is a reasonable expectation of enforcement, or where there is little driver resentment of the specific device. If any of these conditions is not met—for example, where numerous stop signs are used in opposition to major traffic flows or where a turn prohibition is installed and no reasonable (from the driver's viewpoint) alternate route exists—violation of the device can be expected, and should be monitored in the evaluation.

Geometric features, in contrast, physically force or prevent certain driver actions. Although geometric features tend to gain the highest degree of driver compliance independent of enforcement, they also tend to have the most and strongest secondary impacts, both desirable and undesirable. These secondary impacts include traffic diversion to alternate routes, increased circuity of travel, increased emergency response time, and possibly, higher rates of vandalism of the devices. Again, these potential impacts should be examined in the evaluation.

Timing of the Evaluation

In conducting the evaluation, three to six months after implementation should be allowed before "after" data are measured. This gives residents and motorists time to become familiar with the traffic controls and make adjustments. With this interval, the "after" measures will be of *stabilized reactions* rather than *first-encounter responses.* For this same reason, three to six months would appear to be a reasonable period for "experimentation" with devices, or for deciding whether to make "temporary" installation permanent.

This focus of the formal evaluation on stabilized, long-term effects is not to suggest that first-encounter responses and early reactions should be ignored. In fact, these should be carefully observed from the start so that countermeasures to any serious safety problems or obvious defects can be quickly implemented.

While minor adjustments as a result of early surveillance findings often are made during the period immediately after implementation, a commitment should be made to a specific evaluation period before initiating major changes in the scheme. This allows time for traffic and residents to adjust patterns, and if necessary for tempers to cool, and permits evaluation to be based on longer-term performance rather than initial reactions.

Community Involvement in Evaluation

Public inputs to the evaluation are obtained by continuing an active community involvement process. The public can be helpful in providing feedback on their perceptions of how well the plan is working, details of problems, possibilities for improvement, and any aspect overlooked in the initial planning process.

To evaluate in detail the acceptability—both positive and negative—of the project usually requires a more structured approach in the form of a survey or special neighborhood meetings where questions and reactions can easily be focused and addressed to all concerned groups and individuals. Other useful steps are to widely publicize a telephone number for phoning in comments (and/or an address to which written comments should be directed), and to conduct interviews with key informants such as the leaders of neighborhood groups, fire and police representatives, and business associations. For major installations, it also may be appropriate to hold a hearing or workshop before the jurisdiction's planning or transportation commission or even the city council.

Modification of a Neighborhood Traffic Plan

Two kinds of modifications should be distinguished: minor adjustments made in response to identified problems with specific installations or devices, and larger-scale changes to the traffic control concept or approach based on the results of an evaluation of "stable" operations.

Minor adjustments to a neighborhood street's protection plan are a common occurrence. Most such adjustments are physical changes to individual devices (such as moving its placement slightly or altering the signage or markings). Occasionally, a "global" change is made to all devices of a particular type (such as replacing metal posts with reflectorized, breakaway devices). Such changes are intended to improve the devices' operation, eliminate a hazardous condition, or counter deviant driver behavior. Most minor adjustments of either type are undertaken by staff on the basis of their own observations, without any extensive formalized review process.

Observation during the period initially following implementation is critical, in order to identify problems which could easily be eliminated by minor adjustments. It is inevitable that some problems accompany the installation of traffic control devices. Incidents ranging from illegal driving maneuvers to outright vandalism can be expected. Professional staff should be on the scene to observe deviant behavior in first-encounter reactions, to note if any design features are its cause, and to assess whether design modifications can provide a countermeasure to unsafe or purpose-defeating behavior.

Additional police surveillance during the period immediately following installation helps discourage erratic or illegal driving behavior and vandalism. The period of intense first-encounter reaction usually lasts no more than a week or so. After that time, drivers have adjusted their routes sufficiently to avoid the inconvenience caused by the new system.

More important are situations where evaluation suggests that a plan is successful enough that abandonment is not a consideration, but its performance falls short of its intended objectives or has some undesired side effects. Here significant modifications to the plan may need to be considered.

The evaluation stage functions as a needs assessment for such modification. In modifications of this nature, which usually relate to a multidevice plan for a sizable area, one type of device may be substituted for another, some devices may be eliminated entirely, or devices may be added, reoriented, or shifted from one location to another. Normally, this type of modification involves a mini-version of the analytic and participatory processes used in needs assessment, alternatives development, and selection. Because of all that has gone before, the actual activity usually can be compressed in time and scope, though the planning of the modifications should be as thorough and deliberate as the original plan development.

When a plan is deemed to fail irretrievably, "recycling" can occur: the scheme tried is abandoned and the problem is either returned to the alternatives development stage for a fresh approach or one of the previously dismissed planning alternatives is resurrected for implementation. In actual practice, however, when neighborhood traffic control schemes fail, the process usually involves so much controversy that there is no energy or enthusiasm for a "recycling" process.

MAINTENANCE AND ENFORCEMENT ISSUES

Most neighborhood traffic control schemes are designed for the long term—they are intended to be permanent changes in the streetscape. Maintaining the plan, both physically and operationally, therefore requires ongoing attention.

The need to plan and budget for the physical maintenance of neighborhood traffic control devices too often is forgotten. Yet such maintenance is important to the continued effectiveness of the traffic control scheme and to continued public acceptance. Maintenance also can be critical in avoiding liability problems in the case of an accident.

Attention should be given to maintenance issues in designing the plan, as well as in the development of the detailed implementation process and its budget. Key considerations for maintenance include the following:

- There may be high initial costs and continuing budget requirements for the repair or replacement of neighborhood control devices. In early months, vandalism may be a problem (especially with barriers), and the resources to repair and clean up devices should be budgeted. There may also be a need to add such features as signing, marking, and reflectorization to help guide drivers safely around the devices, and to install features such as bollards as countermeasures where some drivers have adopted radical avoidance maneuvers (such as driving across sidewalks and lawns). A contingency fund is one way to assure that such needs can be addressed promptly.
- Increased short- and long-term budget allocations to the departments charged with neighborhood traffic plan maintenance responsibilities should be expected. Usually, there will be an increase in signs, roadway markings, curbs, and other features

that will require regular maintenance. In addition, street sweeping, garbage pick-up, snow removal and similar cleaning functions may require special efforts around barrier-type controls. This should be considered in planning the overall neighborhood traffic control program's budget.

- It is crucial that neighborhood traffic control devices be designed to allow convenient maintenance access to in-street utility features such as manholes, gate valves, vaults and pull-boxes, and catchbasins.

- Landscaping needs care, especially in the early months after planting, but also over the longer term. It could be maintained by public employees, or responsibility could be delegated to neighborhood associations (or even left as a resident responsibility with no formal delegation). Public assumption of landscaping maintenance, while involving some continuing operations costs, assures consistent professional-quality work, avoids potential neglect and disputes among neighbors, and avoids potential adverse liability issues associated with private citizens performing maintenance work on public property. Reliance on private resources, on the other hand, may engender a sense of neighborhood pride and lead to more luxuriant landscape programs than the public jurisdiction would be inclined to fund.

While the physical maintenance of the plan probably will require the greater amount of attention, in a broader sense, maintenance of a neighborhood traffic control plan also requires continued driver respect and public support. Driver respect for the plan may wane if, after the first few months, its enforcement becomes lax; public support, in turn, may fade if lax enforcement—or major changes in other conditions—render the plan ineffective or produce undesirable consequences.

Technical staff responsible for the traffic management plan should periodically observe its operation, looking not only for physical repair needs but also for evidence of violations (such as parking in front of barrier openings intended to be traversable, tire marks indicating regular vehicle passage through emergency vehicle openings, and so forth). Contact with neighborhood groups and residents should be made, in order to learn of any problems they may have noted. It may be necessary to remind police officials of the need for enforcement, or to ask them to do "focused enforcement" in areas that appear to have a high violation rate.

Other changes that may alter the effectiveness or the desirability of a particular traffic management plan also should be considered periodically. In particular, changes in land use within or near the residential area may increase traffic levels or shift travel patterns, with the result that certain control devices are no longer as effective as they once were. For example, a "No Left Turn" sign may deter motorists from violations at a moderately busy traffic signal, but may no longer work if congestion at the signal grows substantially; or a permit parking program which allows two-hour parking by nonresidents may no longer assure on-street spaces for residents if retail uses greatly expand in nearby areas. Staff should try to anticipate such growth-related effects and prevent problems from occurring by proposing appropriate modifications to traffic control plans. Retiming of the signal or restriping of the signalized intersection to increase capacity may be options in the first example; eliminating nonresident parking except by permit or installing controls that deter nonresident entry into the neighborhood may be options in the second.

Occasionally, residents will change their minds about the desirability of a particular neighborhood traffic control device or feature. For instance, they may decide, after an initial period of enthusiasm for traffic control, that increased circuity is more annoying than was the amount of traffic removed from their street. Staff usually learn of such changes in attitude when a group of residents requests removal of the traffic control.

Usually, such removal should be done only after careful analysis and public notice. First, in some cases the device or feature may be part of an areawide control scheme, and its removal could have wide-ranging impacts, perhaps reducing the effectiveness of the controls used elsewhere. Second, it may turn out that the negative aspects of a particular control element could be reduced or eliminated by making a few modifications to it rather than removing it; such a solution may in fact be preferred by residents, once it is brought to their attention. Third, removal of a traffic control may entail substantial costs, especially if it in turn requires changes elsewhere in order to preserve the effectiveness of the remaining elements of the plan. And finally, it may be that only a portion of the residents favor the change; others may strongly wish to retain the contested control.

Public involvement in such cases could take the form of a hearing or workshop, perhaps held by the local planning commission or traffic committee. Documentation of the issues raised, the alternatives considered, their likely impacts, and the views of all affected interests is as important to support a decision to abandon a plan or a portion of it as it is in a plan's initial development.

7

Prospects

Residents all over the world value a high quality of life in their neighborhoods. Traffic is considered one of the main threats to that quality of life, especially when the traffic is noisy and moves at high speeds in comparison to pedestrians and cyclists. A major concern is the safety of children walking, biking, or playing on sidewalks or in the street, and of elderly and handicapped persons. Other concerns include the adverse effects of heavy traffic on the home through intrusion of noise, vibrations, and fumes, and on community cohesion, and the poor aesthetics of a neighborhood in which cars and roadways dominate.

Citizens' desires for more livable neighborhoods not burdened with traffic problems reflect changing attitudes about the role of the automobile, attitudes which favor traffic management rather than continual accommodation. Residents are increasingly demanding that traffic in their neighborhoods be limited and controlled. These same residents, on the other hand, want a street layout which permits them to use their motor vehicles right to and from their front doors. They want their purchased goods delivered and their garbage collected at the same doorsteps. Fast access for fire trucks and ambulances is desired, as are the services of transit vehicles and police patrol cars.

In planning and designing future neighborhoods, it will be a major challenge for urban planners and traffic engineers to balance these conflicting requirements. Street layout will need to be carefully designed and coordinated with land use plans, with the goal of preventing through traffic while at the same time providing good access for residents and the services enumerated earlier, and safe routes for pedestrians and cyclists. An important part of a neighborhood plan's ultimate success or failure will depend on the location of playgrounds, schools, homes for the elderly, and other sensitive land uses, as well as neighborhood-serving shops and other traffic generators.

Traffic engineering and design measures can strengthen a sound planning concept with regard to these requirements. But very seldom can such measures entirely correct a planning concept that has neglected these requirements.

A much greater challenge is presented by the need to correct the traffic problems that plague existing neighborhoods. In the past, too many neighborhoods were built without any thought about the possible hindrance and hazards caused by traffic. With almost no possibilities to relocate traffic generators, improvements have to be made by traffic engineering and urban design methods, both in the neighborhoods and on arterial streets better suited to carrying traffic loads.

In such a situation much imagination and ingenuity is required to design a workable compromise between conflicting interests. To complicate matters further, real problems caused by a proposed redesign of the street pattern of an existing neighborhood will be mixed with imaginary problems. For example, most retail merchants sincerely believe that a flow of traffic in front of their premises is beneficial for their business, even when it is impossible to park a car within an acceptable walking distance of their shops. Emotional involvement of some of the residents will also be the cause of many unfounded opinions, which will often be hard to overcome.

Nevertheless, the redesign of many older neighborhoods will be a necessity to prevent the deterioration of the urban environment. The sheer size of the urban problem is such that, in the vast majority of existing neighborhoods, only very limited means will be available for the improvement of the street environment. This again increases the challenge for those persons who are responsible for the redesign, because poorly designed or unattractive solutions may actually speed up the process of urban decline instead of postponing or reversing it.

This book has attempted to stimulate the reader to think along these lines of planning and action. A number of physical design concepts and examples have been presented, and a range of traffic control devices has been discussed. But the variety available is limited only by the imagination of urban planners and engineers who will need to educate themselves about neighborhood traffic problems, their causes, and their solutions. They need to work together, with planners recognizing the implications of land use decisions and layouts on traffic levels and patterns, and engineers recognizing the multiple uses of streets and the importance of residential satisfaction to a healthy urban environment. The involvement of neighborhood residents in the planning, design, and implementation of traffic management strategies cannot be stressed too highly; only with such cooperation can a plan become accepted and successful.

Changing public attitudes point to the new objectives for street design and traffic control. The conclusion is reached that the goal of expeditious movement of large quantities of vehicles must be applied only to a clearly defined arterial network. The basic needs of urban society do not require that this be the standard by which the performance of local streets in residential areas are judged. On these neighborhood streets the task will be to eliminate traffic unrelated to the neighborhood, to slow down the traffic that still uses the streets, and to manage the parking of vehicles in an attractive and unobtrusive manner. Or, in other words, the challenge is to restore the human scale in the residential neighborhoods of our cities.

APPENDIX

Bibliography: Workable Streets/Livable Neighborhoods

This bibliography of literature relating to traffic management was first prepared by Judith Lamare, Ph.D., in 1982 for the Workable Streets/Livable Neighborhoods Planning Conference organized by the University of California, Institute of Transportation Studies, with a host of cosponsoring organizations. At that time Judith Lamare was Transportation and Urban Development Policy Specialist for the the California Senate Office of Research, and Senator Nicholas E. Petris, Oakland, sponsored the initial effort. Judith Lamare left the California Senate staff in 1983, and since that time the bibliography has been expanded substantially as a consequence of construction of a working database on transportation system management studies maintained to support commercial consulting work in the area of traffic reduction. The items in this bibliography have been selected from that database.

This bibliography brings together references to research studies pertaining to the relationship between the transportation system and the neighborhood environment, and most specifically studies of transportation system management techniques relevant to neighborhood design. The purpose of organizing and publishing this bibliography is to provide technical assistance to those who are confronted with the problems of traffic management in the neighborhood environment discussed in this book.

Following an introductory list on "Neighborhood Transportation Management", this bibliography contains three sections. The first section is titled "Transportation System Management" and contains references to current and proposed policies and practices that are used in urban environments to manage traffic, given existing infrastructure and land uses. This section includes both general works and highly specialized studies of specific techniques. Categories include auto free zones (traffic restraints), bicycles, parking, pedestrian measures, pools, transit, and telecommuting and work hours management. Studies dealing specifically with the work or commercial environment have been excluded.

The second section looks at technology evolution. The section on the evolution of transportation technology includes works on how transportation system changes have occurred in the past, and what is going on at present to indicate emergent technologies. These works remind us that we live in a dynamic world in which opportunities for new solutions to old problems may exist. Alternative fuels, vehicles, and technology evolution are part of neighborhood design for the future.

The third section of the bibliography contains references to selected works that deal with the relationships between the transportation system and its urban environment. Included are studies of the land use and air quality correlates of transportation system issues.

Further questions about this bibliography or the database from which it is derived, may be directed to Judith Lamare, Ph.D., Consulting Political Scientist, 1823 Eleventh Street, Sacramento, California 95814, (916) 447-4956.

NEIGHBORHOOD TRANSPORTATION MANAGEMENT

APPLEYARD, DONALD. *State of the Art; Residential Traffic Management* Washington, D.C.: Federal Highway Administration, 1980.

Bucks County Planning Commission. *Performance Streets: A Concept and Model Standards for Residential Streets*. Doylestown, PA: Bucks County Planning Commission, 1980.

GARRISON, WILLIAM L. *Prospects for Neighborhood Cars*. Berkeley, CA: Institute of Transportation Studies, University of California, 1977.

Institute of Transportation Engineers. *Proceedings: International Symposium on Neighborhood Traffic Restraints*. Washington, D.C.: Institute of Transportation Engineers, 1981.

LAMARE, JUDITH. "The Local Traffic Management Decisionmaking Process in California." Sacramento, CA: California Senate Office of Research, 1982.

LIPINSKI, M.E. "Neighborhood Traffic Controls." *Transportation Engineering Journal*, January 1979.

MALINSKY, G.L. *Residential Neighborhood Barricades—The Ottawa Experiences*. Washington, D.C.: Institute of Transportation Engineers, 1976.

MARCONI, WILLIAM. "Speed Control Measures in Residential Areas." *Traffic Engineering*, March 1977.

MAY, A.D., and EASA, S.M. *Assessing Traffic Management Strategies in Residential Neighborhoods*. Washington, D.C.: Transportation Research Board, 1981.

ORLOB, LLOYD C. "Traffic Diversions for Better Neighborhoods." *Traffic Engineering*, July 1975.

ORLOB, LLOYD C. "Treatment of Through Traffic in Residential Areas." Richmond, B.C., Canada: Western Canada Traffic Association Conference, September 16, 1977.

Public Technology, Inc. *Neighborhood Traffic Controls*. Washington, D.C.: U.S. Department of Transportation, 1980.

SIMKOWITZ, HOWARD, et al. *The Restraint of the Automobile in American Residential Neighborhoods*. Washington, D.C.: Urban Mass Transportation Administration, 1978.

SMITH, DANIEL T., et al. *State of the Art: Residential Traffic Management*. Washington, D.C.: Federal Highway Administration, 1980.

The Urban Consortium. *The Impacts of Traffic on Residential Area Growth Management and Transportation*. Washington D.C.: Public Technologies, Inc., 1982.

WELKE, R.C., and KEIM, W.A. "Residential Traffic Controls." 46th Annual Meeting Compendium of Technical Papers. Arlington, VA: Institute of Transportation Engineers, August 1976.

WYNNE, GEORGE. *Traffic Restraints in Residential Neighborhoods*. New Brunswick, NJ: Transaction Books, 1980.

TRANSPORTATION SYSTEM MANAGEMENT

Aegis Transportation Information Systems. *Videotex Transportation and Energy Conservation.* Governer's Conference. Honolulu, HI: State of Hawaii Department of Planning, January 31, 1984.

Auto Club of Southern California. *Traffic Engineering for Small Cities and Counties.* Los Angeles, CA: Auto Club of Southern California, 1976.

"Better Towns with Less Traffic." *OECD Observer,* no. 75, May 1975, p. 31.

BOLGER, F.T. *Studies of High Activity Density Area Circulation Systems: A Bibliography.* Cambridge, MA: MIT Urban Systems Laboratory, 1972.

BUCHANAN, COLIN. *Traffic in Towns: A Study of the Long-Term Problems of Traffic in Urban Areas.* London, England: Her Majesty's Stationery Office, 1963.

BURKE, J.W., et al. *PARKLOT: A Computerized Analysis Tool for Developing Ridesharing Facilities and Services.* ConnDOT staff paper. Hartford, CT: Connecticut Department of Transportation,

Caltrans. *Guidelines for Ridesharing Facilities.* Sacramento, CA: California Department of Transportation, 1984.

Crain and Associates. *TSM Planning Guidelines for California.* Division of Transportation Planning. Sacramento, CA: California Department of Transportation, 1982.

CRYER, A.J. *Management of Traffic to the Aid of Urban Environment.* Westminister, London: Westminister City Council, 1970.

EAGER, WILLIAM. "Innovative Approaches to Transportation for Growing Areas." *Urban Land,* vol. 43, no. 7, July 1984.

EASA, SAID M., et al. *Traffic Management of Dense Networks. Vol. I—Analysis of Traffic Operations in Residential and Downtown Areas.* Berkeley, CA: Institute of Transportation Studies, University of California, 1980.

ELLIS, RAYMOND, et al. *Impacts of Transportation Control Plans on Travel Behavior in San Diego.* San Diego, CA: 1984.

Federal Highway Administration. *Measures of Effectiveness of TSM Strategies.* Washington, D.C.: Federal Highway Administration, 1981.

Federal Highway Administration. *Traveler Response to Transportation System Changes.* Washington, D.C.: Federal Highway Administration, 1981.

FLYNN, SYDWELL. *The STAR Project: Short-Term Auto Rental—An Enhancement for Large Apartment Complexes.* Los Altos, CA: Crain and Associates, 1982.

GCA Technology Division. *Measuring the Effectiveness of Transportation Controls.* Washington, D.C.: U.S. Environmental Protection Agency, 1975.

GILBERT, KEITH. *Transportation System Management: Handbook of Manual Analysis Techniques for Transit Strategies (Dallas-Fort Worth).* McLean, VA: U.S. Department of Transportation, Urban Mass Transportation Administration, 1981.

KING, R.H. and EAGLAND, R.M. "Gloucester's Traffic and Transport Plan." *Traffic Engineering and Control,* vol. 12, no. 9, January 1971, p. 456-59.

League of California Cities. *California Local Elected Officials' Guide to Transportation System Management.* Sacramento, CA: California Energy Commission, 1981.

LISCO, T.E. *The Value of Commuters' Travel Time—A Study in Urban Transportation.* Chicago: Department of Economics, University of Chicago, 1967.

MAY, A.D., and CORTELYOU, C. *Transportation Systems Management: TSM–type Projects in the United States.* Institute of Transportation Studies. Berkeley, CA: University of California, 1979.

Metropolitan Transportation Commission. *Traffic Mitigation Reference Guide.* Oakland, CA: Metropolitan Transportation Commission, 1985.

MIDGLEY, PETER. "The Art of Traffic Management." *The Urban Edge,* vol. 4, no. 2, February 1980.

MIKOLOWSKY, WILLIAM, et al. "The Effectiveness of Near-Term Tactics for Reducing Vehicle Miles Traveled: A Case Study of the Los Angeles Region." Santa Monica, CA: Rand Corporation, 1974.

Nottingham City Council. *Zones and Collars—Suggested Traffic Control Measures for Peak Travel in Nottingham.* Nottingham, England: The Publicity and Information Department, 1973.

Organisation for Economic Co-Operation and Development. "Overview of 16 City Case Studies." Paris, France: *Urban Transport and Environmental Symposium,* vol. 11, no. 8, July 1979.

Organisation for Economic Co-Operation and Development. *Proceedings*. Symposium on Techniques of Improving Urban Conditions by Restraint of Road Traffic. Paris, France: Organisation for Economic Co-Operation and Development, October 25–27, 1973.

Organisation for Economic Co-Operation and Development. *Reducing Vehicle Travel in Critical Urban Locations*. Paris, France: Organisation for Economic Co-Operation and Development, 1971.

Organisation for Economic Co-Operation and Development. *Integrated Urban Traffic Management: A Report*. Paris, France: Organisation for Economic Co-Operation and Development, 1977.

Organisation for Economic Co-Operation and Development. *Managing Transport: Managing of Transport System to Improve the Urban Environment*. Paris, France: Organisation for Economic Co-Operation and Development, 1979.

Organisation for Economic Co-Operation and Development. *Transportation Requirements for Urban Communities: Planning for Personal Travel*. Paris, France: Organisation for Economic Co-Operation and Development, 1977.

ORSKI, KENNETH C. "Transportation Planning as if People Mattered." *Practical Planner,* vol. 9, no. 1, March 1979.

Peat, Marwick, Mitchell & Co. and KLD Associates, Inc. *Network Flow Simulation for Urban Traffic Control System—Phase 2*. Technical Report No. FHWA-RD-73-83. Washington, D.C.: U.S. Department of Transportation, Federal Highway Administration, Offices of Research and Development, 1973.

PIVNIK, SHELDON I. *Liability in Traffic and Highway Operations*. Chicago: No. 3295, paper prepared for ASCE Convention Exposition, October 16, 1978.

Public Technology, Inc. *Center City Environment and Transportation: Transportation Innovations in Five European Cities*. Washington, D.C.: U.S. Department of Transportation, UMTA, and Office of the Secretary of Urban Consortium for Technology Initiatives, 1980.

REGISTER, RICHARD. *Guidelines and Principles—Elements of Ecological Cities*. Sonoma, CA: Proceedings of the Solar Cities Workshop, August 3-10, 1980.

ROARK, JOHN. *Experiences in Transportation System Management*. Washington, D.C.: National Cooperative Highway Research Program Synthesis of Highway Practice 81, Transportation Research Board, 1981.

ROBINSON, FERROL O., and BARTON-ASCHMAN ASSOC., INC. *Feasibility of Demand Incentives for Nonmotorized Travel: Final Report*. Washington, D.C.: Federal Highway Administration, 1981.

SALVUCCI, et al. *Transportation Energy Contingency Strategies: Transit, Paratransit, and Ridesharing, Parts 1 and 2*. Washington, D.C.: Federal Highway Administration, 1980.

Schwartz & Connolly, Inc. *Innovative and Successful Transportation Control Measures*. Sacramento, CA: Air Resources Board, 1980.

SIMKOWITZ, HOWARD J. *Innovations in Urban Transportation in Europe and their Transferability to the United States*. Washington, D.C.: Transportation Systems Center, 1980.

SIMKOWITZ, HOWARD J. *Transportation System Management, Urban Transportation for the Eighties*. Washington, D.C.: Federal Highway Administration, 1980.

SPARROW, T.F., and DOHERTY, M.J. *The Mobility Enterprise: A New Transportation Concept*. West Lafayette, IN: Automative Transportation Center, Purdue University, 1982.

THOMAS, J.M. *Methods of Traffic Limitation in Urban Areas*. Paris, France: Organisation for Economic Co-Operation and Development, 1972.

THOMAS, T.C. "The Value of Time for Passenger Cars: An Experimental Study of Commuters' Values." *Stanford Research Institute Journal.* October 1968.

Transportation Research Board. *Implementing Packages of Congestion-Reducing Techniques: Strategies for Dealing with Institutional Problems of Cooperative Programs*. Washington, D.C.: Transportation Research Board, 1979.

Transportation Research Board. *Research Problem Statements: Transportation System Management*. Washington, D.C.: Transportation Research Board, 1982.

Transportation Research Board. *Transportation System Management in 1980: The State of the Art and Future Directions*. Washington, D.C.: Urban Mass Transportation Administration, 1980.

Transportation Research Board. *Travel Impacts of TSM Actions*. Washington, D.C.: Transportation Research Board, 1980.

Transportation System Center. *A Comparison of Methods for Evaluating Urban Transportation Alternatives.* Cambridge, MA: U.S. Department of Transportation, 1974.

TWICHELL, JON, and KLEIN, NANCY. *TSM/Ridesharing/Commute Alternatives: A State-of-the-Art Review.* Washington, D.C.: Transportation Research Board, 1982.

Urbitran Assoc., Inc. *Transportation Systems Management: Implementation and Impacts.* Washington, D.C.: U.S. Department of Transportation, Urban Mass Transportation Administration, 1982.

Voorhees, Alan M., and Associates, Inc. *Seven Cities: Estimated TSM Actions Impacts in Relation to Goals and to City Characteristics.* Washington, D.C.: Transportation Research Board, 1979.

Voorhees, Alan M. and Associates, Inc. *Short-Range Planning for Reducing Vehicle Miles of Travel in the SCAG Region.* Los Angeles, CA: June 1974.

AUTO FREE ZONES

BALDWIN, GREG, et al. *Survey of American and European Auto-Free Zones.* Portland, OR: Skidmore, Owings, and Merrill, 1975.

Barton-Aschman Associates, Inc. "Auto Free Zones: A Methodology for Their Planning and Implementation." *Action Plan for Improvements in Transportation Systems in Large U.S. Metropolitan Areas.* Washington, D.C.: U.S. Department of Transportation, Office of the Secretary, 1972.

ELMBERG, CURT M. "The Gothenberg Traffic Restraint Scheme." *Transportation.* Amsterdam, Netherlands: Elsevier Publishing Co., 1979.

HAWTHORN, GARY, and MANHEIM, MARVIN L. "Guidelines for Implementing a Comprehensive Policy of Traffic Restraints." *"Better Towns with Less Traffic."* Paris, France: Organisation for Economic Co-Operation and Development, April 14, 1975.

HERALD, WILLIAM S., and Alan M. Voorhees and Associates. "Auto Restricted Zone/Multi-User Vehicle System Study: Background and Feasibility." *Department of Transportation Report,* vol. 1. no. DOT-TSC-1057. Washington, D.C.: Department of Transportation, 1977.

KLEIN, NORMAN. "Auto Free Zones: Giving Cities Back to People." *City,* March 1972, p. 45-52.

KLEIN, NORMAN. "Auto Free Zones: Giving Cities Back to People." *Ekistics,* vol. 37, no. 19, February 1974.

KUHNEMANN, JORG, and WITHERSPOON, ROBERT. *Traffic-Free Zones in German Cities.* Paris, France: Organisation for Economic Co-Operation and Development, 1972.

LANE, R. *Some Effects of Excluding Traffic from an Existing Residential Area.* 12th International Study Week—Traffic Engineering and Safety. Belgrade, Yugoslavia, September 27, 1974.

LIEBERMANN, WILLIAM. "Environmental Implications of Auto-Free Zones." Draft paper submitted to the Highway Research Board, Washington, D.C., August 24, 1973.

Northeastern University. *The Impacts of an Auto-Free Zone on an Urban Metropolitan Area: Methodology and Case Study.* Boston: Department of Transportation, April 1, 1974.

Organisation for Economic Co-Operation and Development. *Policy Towards the Creation of Vehicle-Free Areas in Cities.* Paris, France: Organisation for Economic Co-Operation and Development, 1972.

Organisation for Economic Co-Operation and Development. *Streets for People.* Paris, France: Organisation for Economic Co-Operation and Development, 1974.

ORSKI, K.C., "Car Free Zones and Traffic Restraints: Tools of Environmental Management". *Highway Research Record no. 406.* Washington, D.C.: Highway Research Board, 1972.

U.S. Department of Transportation. *Auto in the City.* Washington, D.C.: U.S. Conference of Mayors, 1980.

VISKOVICH, BERT J. *Street Closures.* Traffic Study Report Involving Willistors Park Subdivision, Cold Harbor Avenue and Vicksburg Drive. Cupertino, CA: City of Cupertino Public Works Department, January 20, 1975.

WATSON, PETER L., and HOLLAND, EDWARD P. *Relieving Traffic Congestion: The Singapore Area License Scheme.* Washington, D.C.: World Bank, June, 1978.

WIGAN, M.R. "Traffic Restraint as a Transport Planning Policy I: A Framework for Analysis." *Environment and Planning,* vol. 6, August 1974, p. 565-601.

WITHERSPOON, ROBERT. *Motor Vehicle Restraint Manual.* Study Group on Motor Vehicle Restraint Evaluation. Paris, France: Organisation for Economic Co-Operation and Development, January 12, 1973.

BICYCLES

DEAKIN, ELIZABETH. *Utilitarian Cycling: A Case Study of the Bay Area and Assessment of the Market for Commute Cycling.* Berkeley, CA: Institute of Transportation Studies, University of California, 1985.

LOTT, DONNA Y., et al. *Bikeway Usage and Design.* Washington, D.C.: Federal Highway Administration, January 1975.

PINSOF, SUSAN. *Transportation Control Measures Analysis: Bicycle Facilities. Transportation Research Record 847.* Washington, D.C.: Transportation Research Board, 1982.

Sacramento Area Council of Governments. *Bicycle Access to Transit Feasibility Study.* Sacramento, CA: Sacramento Area Council of Governments, May 1982.

State of California, Department of Transportation. *Bikeway Planning and Design: Section 7-1000 of Highway Design Manual.* Sacramento, CA: General Services Publication Section, 1983.

Transportation Research Board. *The Bicycle as a Transportation Mode.* Washington, D.C.: Transportation Research Board, 1976.

Transportation Research Board. *Pedestrian Controls, Bicycle Facilities, Driver Research, and System Safety.* Washington, D.C.: Transportation Research Board, 1977.

WILLIAMS, JOHN, et al. *The University of Montana Cyclist's Survival Guide.* Missoula, MN: University of Montana, Safety and Security Division, 1986.

PARKING

City of Los Angeles. *Parking Management Plan.* Los Angeles, CA: Los Angeles Department of Transportation, 1983.

City of Seattle. *Providence Residential Parking Zone—Six-Month Project Report.* Seattle, WA: City of Seattle, 1981.

CURRY, DAVID; MARTIN, ANNE; and Crain and Associates. "City of Los Angeles Parking Management Ordinance." Washington, D.C.: Transportation Research Board, 1985.

DEMETSKY, M.J., and PARKER, M.R., JR. "Role of Parking in Transportation System Management." *Transportation Research Record 682.* Washington, D.C.: Transportation Research Board, 1978.

DI REZO, JOHN, et al. *Study of Parking Management Tactics: Volume II: Overview and Case Studies.* Washington, D.C.: Federal Highway Administration, 1979.

DI RENZO, JOHN, et al. "Evaluation of Current Parking Management Tactics." 59th Annual Meeting. Washington, D.C.: Transportation Research Board, 1980.

ELLIS, R.H. "Parking Management Strategies." *Transportation System Management. Transportation Research Record Special Report 172.* Washington, D.C.: Transportation Research Board, 1977.

HIGGINS, THOMAS. "Information Bulletin on Flexible Parking Requirements." *Information Bulletin.* Washington, D.C.: Public Technology, Inc., July 1981.

JHK & Associates. *Parking Policies Study for Montgomery County, Maryland; Summary Report.* Alexandria, VA: Montgomery County, Maryland, 1982.

KENYON, KAY. "Increasing Mode Split Through Parking Management: A Suburban Success Story." *Transportation Research Record 980.* Washington, D.C.: Transportation Research Board, 1984.

MILLER, G.K., and EVERETT, C.T. *Increasing Commuter Parking Prices: Impacts at Federal Worksites in Metropolitan Washington, D.C.* Washington, D.C.: Urban Institute, 1981.

National Parking Association. *Recommended Zoning Ordinance Provisions for Parking.* Washington, D.C.: National Parking Association, 1981.

Office of Project Planning. *The Development of a Five-Year Commuter Parking Lot Program.* Hartford, CT: Connecticut Department of Transportation, 1980.

OLSSON, MARIE, and MILLER, GERALD. *Parking Discounts and Carpool Formation in Seattle.* Washington, D.C.: The Urban Institute, July 1978.

PARKER, MARTIN, and DEMETSKY, MITCHELL. *Evaluation of Parking Management Strategies for Urban Areas*. Richmond, VA: Virginia Department of Transportation, August 1980.

PICKRELL, DON, and SHOUP, DONALD. "Employer Subsidized Parking and Travel Mode Choice." 59th Annual Meeting paper. Washington, D.C.: Transportation Research Board, 1980.

SCRUGGS, WILLIAM C. "Residential Parking Permit Program in Arlington County, Virginia." *New England Chronicle*, April 23, 1976.

SMITH, STEVEN; TENHOR, STUART; and JHK & Associates. *Model Parking Code Provisions to Encourage Ridesharing and Transit Use, Including a Review of Experience*. Washington, D.C.: Federal Highway Administration, 1983.

PEDESTRIAN MEASURES

American Automobile Association. *Manual on Pedestrian Safety*. American Automobile Association, 1964.

ANTONIOV, JAMES. "Planning for Pedestrians." *Traffic Quarterly*, vol. 25, January 1971, p. 55-71.

BENEPE, BARRY. "Pedestrian in the City." *Traffic Quarterly*, January 1965, p. 28-42.

BIEHL, B.M. *Pedestrian Safety*. Paris, France: Organisation for Economic Co-Operation and Development, 1970.

BREINES, S., and DEAN, W. *The Pedestrian Revolution: Streets Without Cars*. New York, NY: Vintage Press, 1974.

CAMERON, H.M. "Nature and Value of Present Pedestrian Protection Measures." *Proceedings of Australian Safety Week on Road Safety Practices*. Sydney, Australia: Institute of Engineering, 1967.

CONTINI, E. *Emerging Opportunities for the Pedestrian Environment*. Proceedings of the Pedestrian/Bicycle Planning and Design Seminar. Berkeley, CA: University of California, 1973.

FEE, JULIE ANNA. *European Experience in Pedestrian and Bicycle Facilities*. Washington, D.C.: Federal Highway Administration, 1975.

FOOTE, ROBERT S. "Helping Pedestrians in Urban Areas." Papers presented at the 12th International Study Week—Traffic Engineering and Safety. Belgrade, Yugoslavia, September 1974.

FRUIN, JOHN J. *Pedestrian Planning and Design*. Massapequa, NY: Metropolitan Association of Urban Designers and Planners Press, 1971.

FRUIN, JOHN J. "Pedways Verses Highways—The Pedestrian's Rights to Urban Space." Presented at the 51th Annual Meeting of the Highway Research Board. Washington, D.C., 1972.

GANTVOORT, J.T. "Pedestrian Traffic in Town Centers." *Traffic Engineering and Control*, vol. 12, January 1971, p. 454-56.

GARBRECHT, D. *Pedestrian Movement: A Bibliography*. Monticello, IL: Council of Planning Librarians, Oct. 1971.

Highway Research Board. *Pedestrian Protection: Five Reports. Highway Research Record No. 406*. Washington, D.C.: Highway Research Board, 1972.

Highway Research Board. *Pedestrians: Six Reports. Highway Research Record No. 355*. Washington, D.C.: Highway Research Board, 1971.

LEVINSON, HERBERT S. "Pedestrian Circulation Planning: Principles, Procedures, Prototypes." Proceedings of the Pedestrian/Bicycle Planning and Design Seminar, San Francisco, 1972. Berkeley, CA: The Institute of Transportation and Traffic Engineering, University of California, 1973.

LEVINSON, HERBERT S. "Planning for Pedestrians." Paper presented at the annual conference—New England Section, Institute of Traffic Engineers. Nashua, New Hampshire, December 5, 1972.

LUTHMANN, HAROLD. *Pedestrian Zones in Germany*. Köln, Germany: Deutschen Gemeindeverlag—Verlag W. Kohlhammer, 1972.

MARING, G.E. *Pedestrian Travel Characteristics. Highway Research Record No. 406*. Washington, D.C.: Highway Research Board, 1972.

MAUDEP. *Planning, Design, and Implementation of Bicycle and Pedestrian Facilities*. Conference Proceedings. Toronto, July 14, 1977.

PROKOPY, J.C. *Manual for Planning Pedestrian Facilities*. Office of Research and Office of Development, Federal Highway Administration, Implementation Package. Washington, D.C.: U.S. Department of Transportation, 1974.

PUSHKAREV, B. and ZUPAN, J. *Pedestrian Travel Demand. Highway Research Record No. 355*. Washington, D,C.: Highway Research Board 1971.

PUSHKAREV, B. and ZUPAN, J. *Urban Space for Pedestrians*. Cambridge, MA: MIT Press, 1975.

RUDOFSKY, BERNARD. *Streets for People: A Primer for Americans*. New York: NY: Doubleday and Co., 1974.

SCOTT, W.G., and KAGAN, L.S. *A Comparison of Costs and Benefits of Facilities for Pedestrians*. Washington, D.C.: Environmental Design and Control Division, Federal Highway Administration, 1973.

STUART, DARWIN G. "Planning for the Pedestrian." *Journal of the American Institute of Planners*, January 1968, p. 37-41.

TAYLOR, MICHAEL. "Pedestrians and Cyclists." Presented at OECD Urban Transport and the Environment Conference. Paris, France: OECD, July 10, 1979.

TOUGH, JOHN M. and O'FLAHERTY, COLEMAN A. *Passenger Conveyors*. London: Ian Allan, 1971.

U.S. Department of Transportation. *Pedestrian and Bicycle Safety Study*. Washington, D.C.: U.S. Department of Transportation, 1975.

WOOD, A.A. "Foot Streets: Managing the Environment." *Traffic Engineering and Control*, vol. 11, no. 10, February 1970.

TEMPLER, JOHN. *Development of Priority Accessible Networks: Provisions for the Elderly and Handicapped Pedestrians*. Washington, D.C.: U.S. Department of Transportation, Federal Highway Administration, Office of Research and Development, 1980.

POOLS

CARLSON, CANDACE. "Integrating Ridesharing, Flextime and Transit." Seattle/King County Commuter Pool. Chicago: National Ridesharing Conference, August 1982.

Commuter Computer. *The Vanpool Marketing Incentive Demonstration Project*. Los Angeles, CA: Commuter Computer, June 1979.

CURRY, DAVID, et al. *Evaluation of Seattle/King County Commuter Pool Program*. Los Altos, CA: Crain and Associates and National Ridesharing Group, 1981.

HEATON, CARLA, et al. *Comparison of Organizational and Operational Aspects of Four Vanpool Demonstration Projects*. Cambridge, MA: Transportation Systems Center, April 1979.

HEKIMIAN, ALEXANDER, and HERSHEY, WILLIAM. "Personalized Approaches for Ridesharing Projects: Experience of Share-A-Ride in Silver Spring, Maryland." 60th annual meeting paper. Washington, D.C.: Transportation Research Board, January 16, 1981.

MARGOLIN, J.B., and MISCH, M.R. *Incentives and Disincentives for Ridesharing: A Behaviorial Study*. Washington, D.C.: Federal Highway Administration, October 1978.

MISCH, MARION RUTH, et al. "Guidelines for the Use of Vanpools and Carpools as a Transportation System Management Technique." *Report 241*, National Cooperative Highway Research Program. Washington, D.C.: Transportation Research Board, May 1982.

PRATSCH, LEW. *Driver Owned and Operated Vanpool Market Research*. Washington, D.C.: American University, December 1978.

SEN, A.K., et al. *Ride Sharing and Park and Ride: An Assessment of Past Experience and Planning Methods for the Future*. Washington, D.C.: U.S. Department of Transportation, November 1977.

SUHRBIER, *Vanpool Research: State-of-the-Art Review*. Washington, D.C.: Urban Mass Transportation Administration, 1979.

Tri-County Metropolitan Transportation District of Oregon. *Tri-Met Rideshare Project, Final Report for Portland, Oregon Metropolitan Area. Project Period: 1977-79*. Portland, OR, 1979.

U.S. Department of Energy. *New Approaches to Successful Vanpooling: Five Case Studies*. Washington, D.C.: U.S. Department of Energy, 1979.

WAGNER, FREDERICK A. *Evaluation of Carpool Demonstration Projects*. Washington, D.C.: Federal Highway Administration, 1978.

TRANSIT

BEHNKE, WILLIAM J. *Auto Ride: A Low Cost, Dial-A-Ride System*. Tigard, OR: Aegis Systems Corporation, 1982.

BURNS, LAWRENCE D. *Transportation, Temporal, and Spatial Components of Accessibility*. Lexington, MA: Lexington Books, 1970.

GUDAITIS, C.J., and RAJENDRA, J. "The Development of Park-and-Ride Facilities Program in Connecticut." Institute of Transportation Engineering. Washington, D.C.: ITE, 1981.

KOMIVES, BOB. "Why Not Treat Transit Like a Utility?" *Planning*, December 1979.

Orange County Transit District Planning Department. *Design Guidelines for Bus Facilities*. Santa Ana, CA: Orange County Transit District, 1981.

The Metropolitan Transit Commission. *Guidelines for the Design of Transit Related Roadway Improvements*. MTC-TD-83-01. Minneapolis, MN, January 1983.

The Metropolitan Transit Commission. *Model Code for Transit Improvement*. Minneapolis, MN, 1977.

YAZHARI, DAVID. *Transit Facilities Standards Manual*. Oakland, CA: Alameda-Contra Costa Transit District, March 1983.

WORK HOURS AND TELECOMMUTING

California Department of General Services. *Telecommuting: A Pilot Project Plan*. Sacramento, CA: California Department of General Services, 1985.

California Energy Commission. *Telecommunication and Energy: The Energy Conservation Implications for California of Telecommunications Substitutes for Transportation*. Sacramento, CA: California Energy Commission, 1983.

JONES, DAVID W., et al. *Flexible Work Hours: Implications for Travel Behavior and Transport Investment Policy*. Institute of Transportation Studies Research Report 78-4. Berkeley, CA: University of California, Institute of Transportation Studies, 1978.

JONES, DAVID W. "Off Work Early: The Transportation Impacts of Flexible Working Hours." Paper presented at National Ridesharing Conference. Chicago, 1982.

JONES, DAVID W., et al. "Work Rescheduling and Traffic Relief: The Potential of Flex-Time." Public Affairs Report. Berkeley, CA: University of California, Institute of Transportation Studies, 1980.

Southern California Association of Governments (SCAG). *The Telecommuting Phenomenon: Overview and Evaluation*. Los Angeles, CA: Southern California Association of Governments, 1985.

Transportation Research Board. *Alternative Work Schedules: Impacts on Transportation*. Washington, D.C.: National Research Council, 1980.

TECHNOLOGY EVOLUTION

Technology

ABERNATHY, WILLIAM J. *The Production Dilemma Roadblock to Innovation in the Automobile Industry*. Baltimore, MD: Johns Hopkins Press, 1978.

ABERNATHY, W., et al. "Patterns of Industrial Innovation." *Technology Review*, June 1978.

BLOOMFIELD, GERALD. *The World Automotive Industry*. London: David & Charles, 1978.

CHENEA, PAUL, and General Motors Corporation. "Research in Transportation: The Shape of the Future." *National Forum*, vol. 61, no. 1, December 1981, p. 14.

GARRISON, WILLIAM L. *Impacts of Technological Systems on Cities*. Working Paper no. UCB-ITS-WP-80-17. Berkeley, CA: University of California, Institute of Transportation Studies, 1980.

GARRISON, WILLIAM L. *Innovation and the Structure of Transportation Activities*. Innovation in Transportation: Proceedings of the Workshop. Washington, D.C.: National Research Council, 1980.

GARRISON, WILLIAM L. *Renewing the Automotive-Highway System*. Working Paper no. UCB-ITS, WP-80-11. Berkeley, CA: University of California, Institute of Transportation Studies, 1980

GARRISON, WILLIAM L. *The Future of the Automobile*. Berkeley, CA: Institute of Transportation Studies, University of California, 1981.

GARRISON, WILLIAM L. " Nonincremental Change in the Automobile-Highway System." *Transportation Research*. v. 18A, n. 4, July 1984.

HARLEY, CHARLES K. "The Shift from Sailing Ships to Steamships, 1850-1890: A Study in Technological Change and Its Diffusion." *Essays on a Mature Economy: Britain After 1840*, Donald N. McCloskey, ed., London, England: Methuen & Co., Ltd., 1971, p. 215-34.

HARLEY, CHARLES K. "On the Persistence of Old Techniques: The Case of North American Wooden Shipbuildings." *Journal of Economic History*, vol. 33, June 1973, p. 372-98.

MAK, JAMES. "The Persistence of Old Technologies: The Case of Flatboats." *Journal of Economic History*, vol. 33, June 1973, p. 444-51.

NORTH, DOUGLAS. "Sources of Productivity Change in Ocean Shipping, 1600-1850." *Journal of Political Economy*, vol. 76, p. 953-70.

U.S. Office of Technology Assessment. *Changes in the Future Use and Characteristics of the Automobile Transportation System*. U.S. Office of Technology Assessment Report, vol. 1 and vol. 2, no. OTA-T-83 and no. OTA-T-84, Washington, D.C.: 1983.

WARD, J.D. *The Opportunity for Advanced Ground Transportation Systems*. Washington, D.C.: U.S. Department of Transportation, 1979.

WARD, J.D. et al. *Toward 2000: Opportunities in Transportation Evolution*. Washington, D.C.: U.S. Department of Transportation, 1977.

WHITE, LAWRENCE J. *The Automobile Industry Since 1945*. Cambridge, MA: Harvard University Press, 1971.

Fuels and Vehicles

Acurex Corporation. *California's Methanol Program*. Evaluation Report, Vol. 1, Mountain View, CA: California Energy Commission, 1986.

Acurex Corporation. *Report of the Three-Agency Methanol Task Force*. Vol. 1—Executive Summary. Sacramento, CA: Air Resources Board, California Energy Commission, South Coast Air Quality Management District, 1986.

BOS, PETER, et al. *Commercialization of Electric Vehicles: A New Strategic Perspective*. Cupertino, CA: Electric Vehicle Development Corporation, 1984.

BRENNAN, RICHARD. "The Automobile's Endangered Future." *Futurist*, vol. 13, no. 5, October 1979, p. 317.

BUMBY, JAMES. "The Hybrid Electric Vehicle: Development and Future Prospects." *Futures*, October 1978.

CACKETTE, TOM. "Potential for Methanol Use in California and Its Impact on Air Quality." Statement of Tom Cackette, Deputy Executive Officer, California Air Resources Board, before the State of California Energy Resources Conservation and Development Commission, Sacramento, CA: Air Resources Board, 1987.

CALFEE, JOHN F. "Potential Demand for Electric Automobiles." *Forefront: Research in the College of Engineering*. Berkeley, CA: University of California, Institute of Transportation Studies, 1980.

California Energy Commission. *Hearing on Electric Vehicles*. Sacramento, CA: California Energy Commission, 1980.

HAMILTON, WILLIAM. *Electric Automobiles*. New York: McGraw-Hill, 1979.

Howard R. Ross Associates. *Santa Barbara Electric Vehicle Project*. Santa Barbara, CA: Santa Barbara Metropolitan Transit District, 1980.

Jet Propulsion Laboratory. *Why We Should Have a New Engine: An Automobile Power Systems Evaluation*. Pasadena, CA: California Institute of Technology, 1975.

TRANSPORTATION SYSTEM AND THE URBAN ENVIRONMENT

Air Quality

Air Resources Board. "Guidelines for Air Quality Assessments: General Development and Transportation Projects." Sacramento, CA: California Air Resources Board, 1984.

Association of Monterey Bay Area Governments. *Bicycle Commuting and Air Quality*. Monterey, CA: Association of Monterey Bay Area Governments, 1981.

Bay Area Air Quality Management District. *Air Quality and Urban Development: Guidelines for Assessing Impacts of Projects and Plans*. San Francisco, CA: Bay Area Air Quality Management District, 1985.

GCA Corporation. *Transportation Controls to Reduce Motor Vehicle Emissions in Baltimore, Maryland*. Washington, D.C.: U.S. Government Printing Office, 1972.

GCA Technology Division and TRW, Inc. *Transportation Controls to Reduce Motor Vehicle Emissions in Major Metropolitan Areas*. Research Triangle Park, NC: U.S. Environmental Protection Agency, 1972.

HOROWITZ, JOEL L., et al. *Transportation Controls to Reduce Automobile Use and Improve Air Quality in Cities: The Need, the Options, and Effects on Urban Activity*. National Technical Information Services Report no. PB-240-006. Washington, D.C., 1974.

HOROWITZ, JOEL L., et al. "Transportation Controls Are Really Needed in the Air Cleanup Fight." *Environmental Science and Technology*, vol. 8, no. 9, September 1974.

Institute of Public Administration, Teknekron, Inc, and TRW, Inc. *Evaluating Transportation Controls to Reduce Motor Vehicle Emissions in Major Metropolitan Areas*. Washington, D.C.: U.S. Government Printing Office, 1974.

MUSSEN, IRWIN. *Air Quality and Urban Development: Guidelines for Assessing Impacts of Projects and Plans*. San Francisco, CA: Bay Area Air Quality Management District, 1985.

RANDALL, PATRICK. *Assessing the Air Pollution Emission Reductions from Traffic Mitigations*. Sacramento, CA: Technical Support Division, Air Resources Board, 1985.

ROBSON, ARTHUR J. "The Effect of Urban Structure on the Concentration of Pollution." *Futures*, vol. 14, no. 1, February 1977.

RYDELL, PETER C., and COLLINS, D. "Air Pollution and Optimal Urban Form." Paper presented at the 60th anual meeting of the Air Pollution Control Association. Cleveland, OH: Air Pollution Control Association, June 11, 1967.

SEITZ, LEONARD. "Control of Extended Idling." Office of Transportation Planning. Sacramento, CA: California Department of Transportation, February 1986.

South Coast Air Quality Management District. *Air Quality Handbook for Preparing Environmental Impact Reports*. Office of Planning and Analysis. El Monte, CA: South Coast Air Quality Management District, 1987.

U.S. Environmental Protection Agency. *Guidelines for Air Quality Maintenance Planning and Analysis, Vol. 4: Land Use and Transportation Considerations*. Research Triangle Park, NC: U.S. Environmental Protection Agency, 1974.

Voorhees, Alan M., and Associates, et. al. *A Guide for Reducing Automotive Air Pollution*. McLean, VA: Westgate Research Park, 1971.

Land Use

Administration and Management Research Association of New York City. *Transit Station Area Joint Development Strategies for Implementation, Executive Summary*. New York: Urban Mass Transportation Administration, 1976.

Alameda-Contra Costa Transit District. *Guide for Including Public Transit in Land Use Planning*. AC Transit Research and Planning. Oakland, CA: Alameda-Contra Costa Transit District, 1983.

American Planning Association. "Special Issue on Transit and Joint Development." *Planning*, vol. 50, no. 6, June 1984.

American Society of Planning Officials. "The Bumpy Road to Traffic Diversion." *Planning*, April 1977.

Booz, Allen & Hamilton. *The Impact of BART on Land Use and Development Policy*. Oakland, CA: Metropolitan Transportation Commission, 1977.

Carpenter, Jeff. "Lessening Automobile Dependence through Land Use Planning." *Practicing Planner*, vol. 9, no. 1, March 1979.

Cavallero, John P. *Local Streets. Highway Research Special Report 93*. Washington, D.C.: Highway Research Board, 1967.

Cervero, Robert. "Light Rail Transit and Urban Development." *Journal of the American Planning Association*, vol. 50, no. 2, January 1984.

Elmberg, Curt M. "Compartmentation as a Tool to Reduce Traffic Congestion and Improve the Environment." Paper presented at the seventh summer meeting, United States Academy of Engineering Sciences. Jacksonville, FL: Transportation Research Board, August 5, 1974.

Farr, Cheryl. "Urban Infill: Its Potential as a Development Strategy." *Shaping the Local Economy*. Washington, D.C.: Real Estate Research Corporation, 1984.

Hafevik, George, ed. *The Relationship of Land Use and Transportation Planning to Air Quality and Management*. New Brunswick, NJ: Rutgers University Press, 1972.

Hamm, Jeffrey. "Conditioning Building Permits with Ridesharing Mitigation Measures: The Seattle Case." Annual Meeting paper. Washington, D.C.: Transportation Research Board, 1982.

Kendig, L., et al. *Performance Zoning*. Chicago: American Planning Association, 1980.

Lamare, Judith. "Land Use-Transit Coordination Policy: A Challenge to Transit Operators in Growing Urban Areas." Annual Meeting paper. Washington, D.C.: Transportation Research Board, 1986.

Larwin, Thomas F. "Characteristics of Selected New Towns or Planned Communities." *Transportation Engineering*, vol. 47, no. 9, September 1977, p. 40.

Lutin, Jerome, and Markowicz, Bernard. "Estimating the Effects of Residential Joint Development Policies on Rail Transit Ridership." *Transportation Research Record 908*. Washington, D.C.: Transportation Research Board, 1983.

Lutin, Jerome, and Bergan, John P. "Joint Development Prototypes in the Northeast Corridor." *Transportation Quarterly*, vol. 37, no. 1, January 1983.

Myers, Phyllis, et al. *Thinking Small: Transportation's Role in Revitalization*. Washington, D.C.: Urban Mass Transportation Administration, Office of Policy, Budget and Program Development, 1979.

Pickrell, Don, and Shoup, Donald. "Land Use Zoning as Transportation Regulation." 59th Annual Meeting paper. Washington, D.C.: Transportation Research Board, 1980.

Potter, Paul. "Urban Restructuring: One Goal of the New Atlanta Transit System." *Traffic Quarterly*, vol. XXXIII, no. 1, 1979.

Public Technology, Inc. *Growth Management and Transportation*. Urban Consortium Information Bulletin, DOT-I-82-28. Washington, D.C.: Department of Transportation, 1982.

Pushkarev, Boris, and Zupan, Jeffrey. *Public Transportation and Land Use Policy*. Bloomington, IN: Indiana University Press, 1977.

San Francisco Department of City Planning. *Street Liveability Study*. Consultant, Donald Appleyard. San Francisco, CA: June 1970.

Sedway Cooke Associates. "Joint Development." *Urban Land*, vol. 42, no. 7, July 1984.

Southern California Rapid Transit District. *Land Use and Development*. Los Angeles, CA: Southern California Rapid Transit District, 1982.

Southern California Rapid Transit District. *Joint Development and Value Capture in Los Angeles: Local Policy Formation*. Report no. DOT-1-83-23. Washington, D.C.: Urban Mass Transportation Administration, 1983.

St. Louis Community Development Agency. *Hyde Park Restoration Plan*. St. Louis, MO: St. Louis Community Development Agency, 1976.

Stringham, M.G.P. "Travel Behavior Associated with Land Uses Adjacent to Transit Stations." *ITE Journal*, April 1982.

Transportation Research Board. *Coordination of Transportation System Management and Land Use Management.* Washington, D.C.: Transportation Research Board, 1982.

Urban Land Institute. *Joint Development: Making the Real Estate-Transit Connection.* Washington, D.C.: Urban Land Institute, 1979.

WITHERSPOON, ROBERT. "Transit and Urban Economic Development." *Practicing Planner*, vol. 9, no. 1, March 1979.

Index